The Stream Conservation Handbook

The Stream Conservation Handbook

Edited by
J. MICHAEL MIGEL

Introduction by
NATHANIEL P. REED

Illustrations by
DAVE WHITLOCK

WILLIAM FLICK
BEN EAST
J. MICHAEL MIGEL
DONALD ECKER
STANLEY BRYER
MAURICE OTIS
DAVE WHITLOCK
P. VAN GYTENBEEK
ALVIN R. GROVE, JR.

CROWN PUBLISHERS, INC.
NEW YORK

Library of Congress Catalog Card Number: 73–82958

Designed by Ruth Smerechniak

Manufactured in the United States of America
Published simultaneously in Canada by General
Publishing Company Limited

To those anglers who accept the challenge to be surveillant of pollution trespasses, and who will become the honorary stewards of pure, clean streams

CONTENTS

Introduction

NATHANIEL P. REED

An individual who has the opportunity to serve as Assistant Secretary for Fish and Wildlife and Parks in the Department of the Interior is exposed to many scientific reports, to briefings by professional biologists and conservationists, and to correspondence from many people deeply concerned with the environment and more specifically stream conservation. Thus, when Mike Migel asked me to review this book and write the introduction, I was delighted for I either know the authors personally or am aware of their contributions to conservation. I was literally hooked!

As a youth I was privileged to fish for trout and bass both in the East and in the West. Thus began a love affair with streams and rivers—running water. During my early years, I saw too many stream fisheries grossly damaged by concrete monuments constructed in the name of "progress." I witnessed the Atlantic salmon streams of Maine and Canada slowly begin to die due to river pollution, overfishing of the high seas, misused pesticides, and dams that stopped the salmon's historic migration to spawning grounds. I watched the death of smallmouth bass rivers which were dammed, channelized, exposed to acid mine drainage, and polluted by industrial and municipal wastes. It seemed to me that one of the biggest tasks was to educate the American people that these abuses could be halted by an aroused public.

It is extremely rewarding to see the many authors of THE STREAM CONSERVATION HANDBOOK put forth their views on this vital subject. Each author presents his contribution in his own personal style so that *his* philosophy comes through loud and clear. Professional biologists and laymen alike may debate some of the conclusions and recommendations presented here. This is to be expected, in fact warranted, on a vital subject such as this. The ecosystem of a stream is a living thing, ever changing, not only in season but in space and time. Controversies about stream conservation will remain with us always. I view this as a healthy phenomenon, as it makes us find newer and better remedies.

One thing that impresses me about the writers of this book besides the obvious competence in their selected field is that they are all fishermen. I have always believed that if you are to inform someone what is good or bad for him you need the intimate experience of doing it yourself. Those who angle, drift, or paddle a boat down a meandering stream, who pack into wilderness watersheds, not only are knowledgeable of these resources, but they gain a reverence and appreciation of our lands that can never be obtained by other than personal experience. When you have challenged a great fish with a perfectly presented fly, lured him, played him down, and released him to live and perhaps to be caught again, you have experienced emotions of mind and heart that strengthen one's integrity and give a new dimension to life.

Mike Migel talks about us fishermen becoming an endangered species in the first chapter. Beyond this, Interior's Bureau of Sport Fisheries and Wildlife, in its recent publication *Threatened Wildlife of United States,* lists thirty-odd endangered fishes, five of which are trout. Moreover, this doesn't include thirty-five other species of fish that are threatened with extinction or on which information is so scarce that we have to list them as "status—undetermined." Many of these are not sportfish but it still reflects a serious situation—a situation that as an avid fisherman scares me. It is not the anglers who are responsible for the conditions which caused these species to become endangered but water-development programs, careless or thoughtless mining operations, improper timber harvesting, overgrazing of livestock, inadequately conceived real estate development, and poorly executed stream straightening—these are the elements that are expediting the destruction of the stream environment of America.

This book will give you basic insight into a stream's ecology and an understanding of the delicate processes which produce as an end product—wild fish. The battles we fight for stream conservation are also documented throughout the book—both victories and losses. Fortunately victories in the war against the transgressors and the abusers now come more often. The 1969 National Environment Policy Act gave us a new law while we dusted off an old one, the Refuse Act which was enacted before the turn of the century. Both laws when enforced have given needed protection to stream environment. In the past decade we have seen our citizens become greatly involved with environmental groups dedicated to saving and managing the natural resources of our country. With NEPA and the Federal Water Pollution Control Act of 1972, which amended the 1899 Refuse Act, our war with the factions that are ruining our rivers is slowly turning in our favor. If our nation's streams and rivers are to be protected from man's greed, it will be due to an educated and aroused public.

Why fight for a stream's future? Roderick Haig-Brown summed it up eloquently in *A River Never Sleeps:* "I don't know why I fish or why other men fish except that we like it and it makes us think and feel. But I do know that if it were not for the strong, quick life of the rivers, for their sparkle in the sunshine, for the great coldness of them under rain and the feel of them about my legs as I set my feet hard down on rock or sand or gravel, I should fish less often."

Good reading and tight lines!

Note

The Stream Conservation Handbook has been conceived as a primer—and a plea; it is *not* an exhaustive study of all aspects of stream conservation. Some of the chapters inevitably overlap, and in some cases the opinions of one contributor differ slightly from those of another; to have edited out all such repetition and controversy would have been as unworthy of this complex subject as of the integrity of the distinguished contributors. Hopefully, this book will engage you with its profoundly important subject, make you more aware of its particularities and—most significant—*get you involved* in active work on behalf of our rivers and streams.

1

Fishermen—Another Endangered Species?

J. MICHAEL MIGEL

> *To thoroughly enjoy angling, besides fishing for sport, one must learn to become a student of the stream food chains, to strike up a friendship with the banks, the trees, the birds, the environment and, above all, to make a true companion of any waters where one casts a line.*
>
> MICHAEL FROME

1

Izaak Walton and Charles Cotton, in their gentle classic *The Compleat Angler,* advocated a dual approach to the pursuit of freshwater fish: they stated that a memorable part of a day was fishing "fine and far off," but another part, just as important, was the natural grace and tranquillity that can be enjoyed by anybody who approaches, rests nearby, or wades into a fish's habitat. These thoughts started a tradition that has lasted and is a path trod today by all good anglers.

For water, sweet flowing water—clear and cold in the northern reaches, where the trout and salmon fisherman stretches his arm and pleasures his eye, warmer and more turbid in the South, where the bass and bream fisherman boats—is where any fishing adventure begins, is consummated, and lives. And any fisherman, be he man or boy, who has any feelings whatsoever for his sport, carries with him as a personal possession an aching tenderness for some creek, stream, or river. His mind's eye can instantaneously project a run, a pool, or a riff that has yielded him its bounty, or where a lunker has broken him off. An angler's river is his summer home. It is the source of many of his sweetest winter dreams.

My river was the Schoharie and, like those of all fishermen who have haunted a particular stream, my memories could fill a book.

For years, each leaf-budding season, I'd walk the Schoharie's banks, anxious to see what changes the spring spates had carved out. Usually I was contented: most of the old holds, the secret places had endured—slightly different this year, perhaps, but still there.

Then, later, on days of lazy cumulus clouds and sometimes gray-black thunderheads, my stream and I flirted. There were the long, warm twilights when the trout dimpled and took—and other evenings when the soft air was dotted with mayfly spinners, but there were no rises, and fish had to be pounded up.

The close of each season in these mountain altitudes brought frost and shortened days and a painted landscape. I angled in drought-lowered waters with my flies playing hide-and-seek between brown and crimson sailboat leaves. The waters were quiet and breathless as they awaited the coming cold. And finally, each late fall or winter as I passed by on my way to woodcock or grouse covers, or crunched through the snow waiting for the cry of the beagles—"We've got a snowshoe started"—the runs looked dark and blue-black. Often water dripped from the banks, or icicles and icefalls could be seen, their crystalline shapes rainbowed; the equinoctial storms had filled the earth fissures and raised the water table so their flow renewed the Schoharie's level. And I felt the cycle would repeat and repeat, and I'd be the river's willing captive for all my angling years.

But I was wrong.

Several years ago, my river and I began to part company. I still took fish, but they were generally small, and I became restless. The river was different. Like so many other fishermen when they become dissatisfied, I started looking for new waters. I sought no reasons; I looked elsewhere.

From the beginning of my relationship with the Schoharie, my mentor had been Art Flick. Late one evening recently, I told him that day I'd caught a small rise just right at the Green-

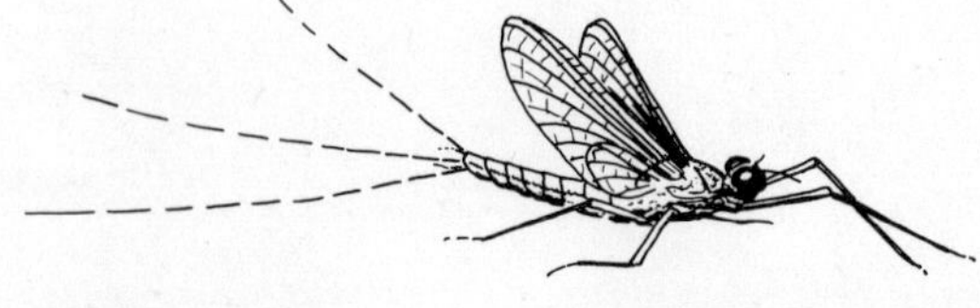

house Pool and had taken and released a number of fish; two were good, the rest stockers, still green. But somehow, the river was definitely not what it once was, I said.

Art looked at me for a long while and then asked, "What did you see?"

Thinking he meant mayflies, I answered, "Early, some Grey Foxes and then later, some *Dorotheas*."

"No," he said, "I didn't mean the flies. Did you happen to see any trash in the river?"

I thought for a long moment or two. "The one fish I kept was taken near an old sunken truck tire. I took a stocker near a waving piece of plastic; from a distance, it looked like a dead eel."

"Did you see any beer cans?" Art persisted. Remembering, I acknowledged I'd seen several. Art changed the subject. "How deep do you think the water was along the far edge of the Greenhouse?"

"Three or four feet," I answered.

"Six or seven years ago, if the river was running at the same height, how deep do you think the water would have been?"

"Eight to ten feet."

"Today what was most of the bottom like?"

"Fine sand, some mud, loose gravel," and, finally catching Art's drift, I kept on, "and years ago it was coarse gravel with rocks sticking up through it."

For two hours, Art went over the history of the Schoharie; part took place before my time, and part I knew. A dam had been built at Prattsville; from there the water could be fed through a tunnel under the mountains to where it junctioned with the Esopus at the Flume. For years the Prattsville reservoir had been deep enough and cold enough for rainbow trout to summer there, and in the spring to migrate up and down the Schoharie. But eventually silt had filled the lake; now it held bass and walleyes.

Though the hurricane of 1954 and several years of drought had hurt it, the river would have come back; but a ski resort had been developed at Hunter. This was good for the business people because the town changed from a summer resort to an all-year-round spa. It was bad for the Schoharie because the developers had erected their apartments and condominiums too close to the river and were using too much of its water. Down the road from Hunter, people had built summer homes along the banks; tourist courts proliferated. Sewage and detergents caused pollution.

Finally, the two-lane road that ran the length of the valley was converted to a superhighway, and during and after its construction, tons of silt and loose gravel washed into the stream. Spring holes were covered, water temperatures rose too high in July and August, and the depths of all the famous pools where the trout could summer—Baseball Pool, Killer Rock, Mosquito Point—grew shallow and offered no sanctuary. There were few holdover or big trout anymore. A great deal of the fishing nowadays was just dodging trash and catching stockers.

Like many stream fishermen, I'd paid lip service to conservation: I read news articles cursorily; I belonged to a sportsmen's group or two; and, if the subject came up while I was with any of my fellow anglers, I'd listen with half an ear.

Art's conversation was a turning point. Bad real estate developments, pollution, unimaginative road building, and siltation —all classic causes of stream degradation—had hurt a river I loved; and it had taken place under my very eyes. I had seen all of it happen. The changes had been gradual, but finally *inescapable.* My only answer had been to look for new waters—partly because big trout had become scarce, but chiefly because the flavor of the river had changed.

Now I sought the root causes.

And I sought better answers.

2

Population growth is probably the single greatest cause of stream destruction.

Up to the end of World War II, the expansion of the population in the United States had been slow but sure. In 1850 there were 25,000,000 citizens; by 1944 there were 145,000,000. But after 1945 there were dramatic changes. Gripped by the idea of expanding markets, postwar prosperity, and rapidly advancing medical and industrial technology, Americans let a fever run rampant in their own country.

They took on a new worship—*growth.* They forgot that growth is not a quantity but a quality; they ceaselessly labored for bigness in their cities, their factories, their stores, their farms, their dwelling places, their automobiles, and even the

An inspiring sight: a stream system in its natural relationship to the rest of the ecology. This section of the Teton River is due to disappear beneath a damsite. *J. D. Roderick, United States Bureau of Reclamation*

size of their families. The population did not soar—it exploded. In 1973, 228,000,000 people live in these United States.

These people must consume vast quantities of industrial products.

They must drink.

They must vacation from the teeming gray megalopolises they inhabit.

They read newspapers, books, and magazines made from trees.

They produce wastes.

Each of these activities, and many more, affects the life of rivers.

And now, with rising incomes and the ease of travel, families are taking to the roads in record numbers. Consider this *one* statistic: In 1970, 6,778,500 people visited Great Smoky Mountains National Park in Tennessee and North Carolina. This number is expected to increase by 20 percent in 1980, when 8,135,200 individuals (more than the total population of many small nations) may come to this one park.

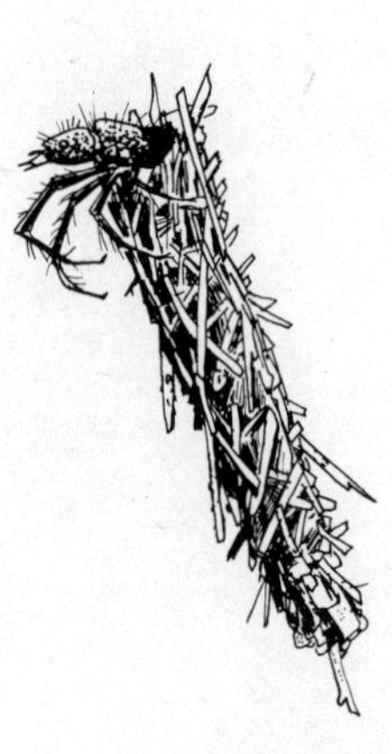

The percentage of fishermen has more than exceeded the general population explosion. Today, one out of every six Americans fishes. Last year 32,383,955 licenses were issued and an accepted figure is that 12½ percent more people fish than buy licenses. With every passing day, fishing intrigues more and more people. The press of civilization, the tension in modern life, makes many men almost desperate to escape their four walls and seek sanctuary by returning to one of the simplicities of their pioneer heritage—fishing.

The ecological demand has become staggering.

Unmonitored per capita consumption of natural products has created awesome problems. Careless destruction of our ecosystems, willful depletion of our resources, and thoughtless industrialization have increased in the last twenty-eight years in direct proportion to an uncaring population.

Recently a group of scientists came together under the banner of the Massachusetts Institute of Technology for a study of "Critical Environment Problems." They coined the phrase, "Ecological Demand" and defined this as the "summation of all man's demands on the environment, such as the extraction of resources and the return of wastes." They predicted it would always keep pace with—or outstrip—population growth.

From pre-Columbian times through the early part of the nineteenth century, approximately three million miles of waterways were part of the United States. During this time, even when small urban communities along the coasts or in river valleys were struggling to establish themselves, the rills and rivers from Canada to Mexico and from the Atlantic to the Pacific ran cyclically, raised and lowered, flash-flooded and froze, cleared and carried silt, but abnormal erosion was retarded by the native flora of the land through which the flow moved. For nature, through evolutionary processes, had induced the growth of grasses, shrubs, and trees whose roots absorbed much of the rain and snowfall and prepared the soil to return the excess water slowly. Marshes and wetlands temporarily cradled overflows. During this period, there were few inhabitants in this vast land, and the "ecological demand" was minimal. But the drums in the evolutionary march beat inexorably and with a slow and measured cadence.

From the early nineteenth century until the end of World War II, the number of people increased, and they slowly and surely spread across the country; first horses, then covered wagons, railroads, and finally cars and trucks carried them. The complexion of the land changed. Much of it was no longer wilderness; industry flourished along the waterways, tremendous amounts of acreage were brought under the plow, but the country was vast enough to absorb all these incursions into its ecology. The population growth was not large enough, its industrialization had not advanced far enough, the "ecological demand" was not great enough for the air, the land, and the waters to be noticeably contaminated.

Now it is otherwise.

One of the factors people disregarded was the vital necessity for human beings, no matter where they live, or where they visit, to have *clean* water. Each person in the United States uses about 1,800 gallons of water per day (150 gallons personally and 1,650 gallons for industrial and power production). In 1970, we, in this country, used 370 billion gallons daily. (The average daily flow of the great Mississippi River is 395 billion gallons daily.) *Clean* water is one of our most precious commodities.

People have deluded themselves with the idea that as the demand grew for more and more billions of gallons of water, it

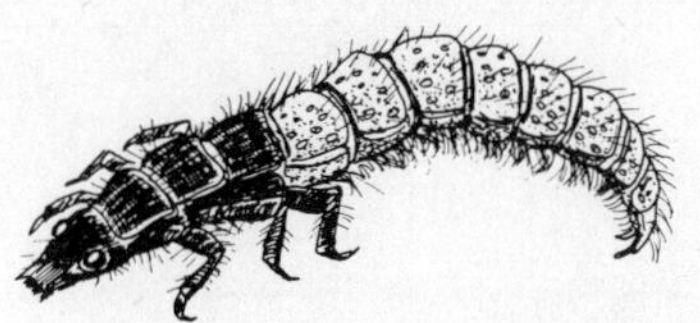

could be obtained easily from distant mountain ranges and drainage areas. Fifteen or twenty years ago, there were many such vast tracts, but today there are few wilderness areas that haven't been affected by pollution or industrial exploitation.

Appearances *are* deceiving. Flying across the United States, one cannot help but be impressed by the vast amount of supposedly "untouched" land. One could get the impression the population explosion is a scare tactic dreamed up by doomsday prophets—that there are thousands of acres and miles of streams still waiting for exploration. Look out of the window again. How many roads crisscross the "virgin" territories? If there is flowing water beneath you, study it. Has the surrounding forest been timbered too severely? Probably. Has mining or strip mining caused erosion? In some areas, definitely. Is the river or stream flooding too much? Is it dried up? Is it dammed? Is it a place where you'd like to fish—or has it already been overdeveloped? People everywhere, and in the urban centers especially, have drawn, and must continue in numbers to draw, their heartbeat from the water and from such natural resources as wood, oil, and gas.

There are no more remote areas; there can be no more complacency. The very plane you are on has been a factor in shortening all distances. There is no place in the United States that can't be visited in a few hours. Our country, which a few years ago seemed so unlimited, has shrunk and in the future will continue to shrink.

While looking to faraway places for clean water, people, driven by their expansion goals, proceeded to run amok and rape whatever brooks, streams, and rivers were near them. No regard whatsoever was paid to flowing water as a living entity. There was no attempt at compromise. There was only the frenetic effort to build, to grow. If trees stood in the way, they were bulldozed; if wetlands could be drained, they were channelized—never mind that in 1971 by government fiat, we paid farmers more than $2.75 billion to idle thirty-seven million acres of surplus cropland or that the Soil Conservation Service estimates that in 1970 one billion tons of topsoil was washed downstream by floods and waters that flow too quickly. If a real estate development could realize higher profits by being erected on a stream bank, tapping the water, and dumping back insufficiently treated sewage, it was so built.

Dump industrial wastes into a river. It will carry them off. Fabricate houses in flood-prone areas. The government, the state, will build levees, or again channelize, to "protect" them. Never mind that these waters will be rushed downstream; build more levees and channels there. Invent more persistent pesticides. When they've done their duty, the rains will wash them away. Sell more detergents. Never mind that some of them kill useful bacteria and accelerate algae growth, which deadens life in water: flush them down. Drop garbage anywhere, anytime. Throw beer and soft drink cans in a brook; it will take only 400 years to recycle an aluminum can, 300 years a tin can, and up to 170 years for plastic.

What happened in a small way to the Schoharie has happened to Goose Creek, the Cuyahoga, the Potomac, and thousands of other rivers and streams. It has happened in every state. By 1970 88.8 percent of all the flowing water in the entire United States was polluted or degraded! Not only are fishermen on the verge of becoming an endangered species, but all human beings in this nation—which only a few years ago boasted about its great number of wild rivers—are at an extreme danger point: their clean water is vanishing.

Yes, ravaging ecological demand has outstripped a vastly growing population. Few people gave any thought to what could be done to compromise the various aspects of such demand so that sane growth could continue while nature was still protected. Fortunately, within the last few years, the few people infected others. All their voices have been rising, until today they are a chorus that *will* be heard. The federal government and the state legislatures have had to do some listening, and bills have been passed. Congress authorized the spending of $7.8 billion for water purification. The people in New York State, by a two-to-one majority, authorized an Environmental Bond Issue —$2.5 billion. With help such as this, the contaminating flood has stopped flooding, *but it is still rising*.

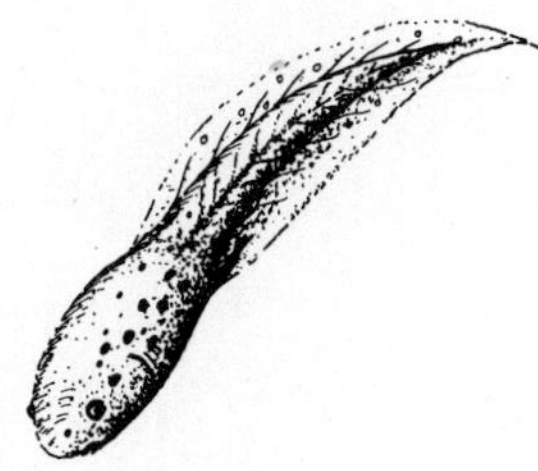

Soon, hopefully, the United States will reach a state of "demographic transition," with births and deaths equalized; but until it does, population growth—the basic problem—will continue. Ecological demand, and hopefully it will be *sane* ecological demand, will keep pace with the numbers of people. Technology will advance, industrialization will grow, building will swell. No one, no foreseeable force, can stop them. But the fisherman, who not only uses water but has his recreation in it, can and must exert his muscle. Not to attempt to stop these forces, for that would be impossible, but to make these forces consider at all times before they make any forward movements, what the effects of their clearing, building, channelization, damming, polluting, and littering will have on the environment, on flowing water, and to force them, if necessary, to back off or to compromise if their efforts are dangerous.

No one can expect the brooks, streams, and rivers of today to be returned to their pristine delights of fifty years ago. Some rivers have been saved, and some have been "brought back." Important work has already been done and more is underway. Awareness has increased. But much more work remains before the frightening curve of destruction will be stopped, so that our flowing water in 1980, 1990, and 2000 will not be the slimy, brown sluiceways forecast today.

3

"Why should you get involved?"

Many conservationists answer this by stating, "If you don't, there will be nothing left for your children." I have children and grandchildren, and I'm concerned for their future. There may be a better appeal.

I'm willing to work so that my descendants can have a place to fish, but of equal and perhaps even greater importance is that I, myself, want to angle five, ten, perhaps even twenty years from now. The way things are going, this may not be possible.

What can you and I do? We are workers, businessmen with families to support. There are professional conservationists who get paid. There are dedicated men in government and in clubs. Television, newspapers, and magazines all now carry stories about saving our streams and rivers. What can a single individual or a group of individuals accomplish?

ZAP is the stuff that beams out of Broom Hilda's finger and used to spray out of Buck Rogers's gun. It makes everything bad disappear. The *Journal-Bulletin* of Providence, Rhode Island, tired of paid consultants, expensive surveys, and political promises, named and publicized a project to clean up one of the filthiest cesspools in the East, the Blackstone River, ZAP. It was the promotion department's contention that individual citizens from Rhode Island and Massachusetts would respond if appealed to. And were they right.

On September 6, 1972, ten thousand concerned citizens turned out. They were a cross section from all walks of life—laborers, professionals, unemployed, managers of construction companies and their equipment operators, military reservists and National Guardsmen, children, students, housewives, Boy Scouts, teachers, Girl Scouts, Sea Bees, secretaries with their bosses, newspaper reporters, and clergymen. At the end of one day ten thousand people had dragged from the bottom of the Blackstone River *ten thousand tons* of junk and trash—two thousand pounds per worker. This was piled along the riverbanks, waiting to be trucked away to the many community dumps ready to receive it. Phase I of ZAP did not change an industrial sewer into a clean fishable river, but it was a start, and it proved what aroused individuals can accomplish. The story was carried in *Field & Stream* magazine, and it is not too much to hope that many of its vast readership will begin their own local projects.

There are thirty-nine million fishermen. If each wants to save his fishing by cleaning and keeping his waters clean, we can. But we must be motivated. We must want to learn what is wrong. We must take time to study our streams, not just fish them; to learn what aid is available, and from what specific sources. Not just one or two but all of us.

This book is designed to give you an insight into problem areas—how to spot them, and what you can do about them. It has three sections. To establish a base, William Flick documents the basic parameters of a normal stream; Ben East then points out the major types of man-made desecrations and other concerned experts give you definitive methods that you, as an individual, can use to do your part to help achieve environmental sanity.

Flowing water conservation is creating a state of harmony between the rills, brooks, streams, and rivers and man. It will be a long time before this simple statement becomes a reality. We will have to learn, as a moral ethic, that water and its sources are not private possessions to be used and abused at anybody's whim. A man's home may be his castle, but he will have to conduct himself in a way that will not destroy or pollute even a small part of any fundamental resource.

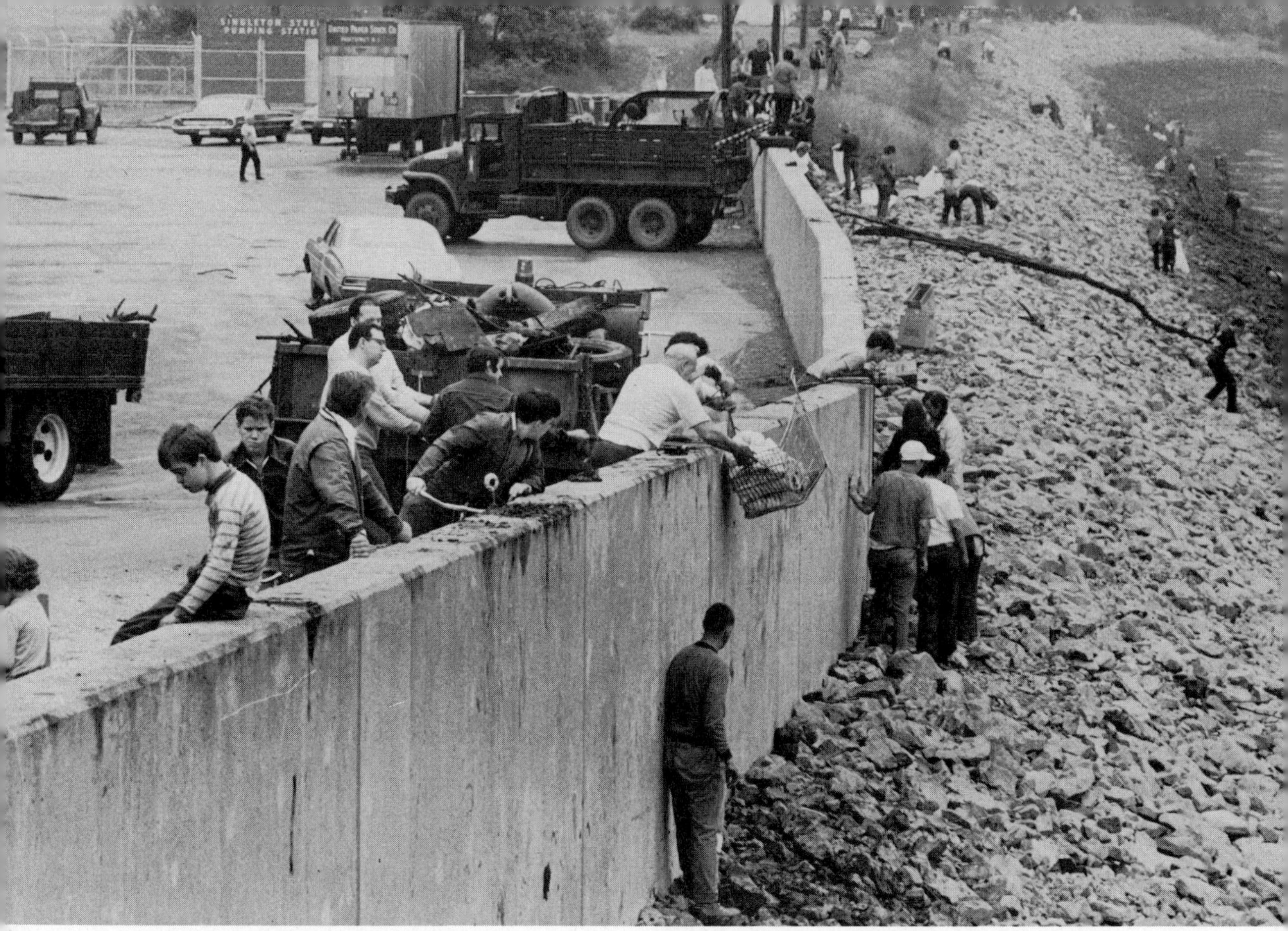

Help for the Blackstone River in Rhode Island came from all directions during ZAP, and it was as effective as it was gratifying. No human being seems to have lost the sense that his home is really the whole environment—and this is the most basic expression of ecological consciousness. That sense, rallied to action, could go a long way toward preserving and reclaiming our streams—and it did here. *Photos by Dick Benjamin, courtesy of The Providence Journal*

One of many scenes along the Blackstone during the ZAP. *Photo by Dick Benjamin, courtesy of The Providence Journal*

No longer can an individual or a developer build on the banks of a stream without providing adequate sewage treatment.

No longer will a farmer be able to use pesticides in excess, if the runoff is a contaminant.

No longer can the refuse of factories be dumped into a river.

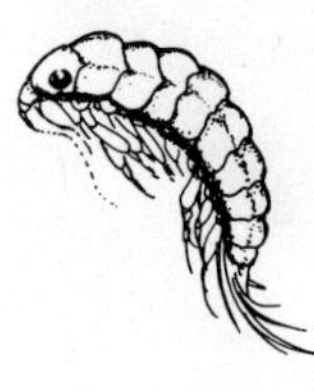

And this will happen, not because of laws, because for the most part there are extant laws, but because people will have a moral credo that states: "Water and air are precious and belong to all the people and no individual has the right to abuse them."

Government must and will play its part by passing laws; it must itself refrain from indiscriminate dam building, clear-cutting, and channelization; it must furnish money, though no amount of money, no matter how broad the tax base, no matter how severe the penalties, can do a job, unless a universal water ethic is developed—a social conscience about water and its environs.

As a capitalistic society, one of our idols remains self-interest. It is wrong to tell a farmer not to pasture his cattle on a too-steep hillside bordering a stream, because of the terrible erosion problems, if it is his only pasture, and if he and his family will go on relief if his dairy herd doesn't get grass. But if there is universal concern, some remedy can be achieved; perhaps a neighbor might help him fence a strip with trees planted to protect the stream. There are *always* solutions. One of the basic tenets of capitalism is the judicious use of prime resources. Nature, the cornerstone of all our lives, is our prime resource.

Eventually, our entire economy must learn to believe in Aldo Leopold's theme: "a land ethic changes the role of Homo Sapiens from conquerors of the land community to plain member and citizen of it." Industry cannot destroy or pollute because in the end, if it does, it will destroy its own marketplace.

How can such an ethic be established? How is *any* moral code established? Eventually it is the will of the majority of the people, but it starts by a few individuals holding to a belief. They talk about it among themselves, with business people, farmers, and builders. They discuss it with their children. They bring it up at PTA meetings and persuade the educators to write or sponsor textbooks that expound this principle. Not in nature classes but in English, history, science, and economics classes. The Leopold tenet—that in today's heavily populated world no individual, no company, has the right to pollute or desecrate any natural resources for selfish reasons—must become a tenet for us all. Above all, if they are fishermen, they set an example by working at streamside conservation themselves; they learn to watch and publicly praise good water-ethic conduct when they see it, and to voice their opposition when an act contravenes the basic principle. You may not feel one person's contribution is worth much, but you, plus thirty-nine million others, are a mighty army. Get in the habit, when you read newspapers and magazines, of keeping a sharp eye out for any national or state conservation bills that are before Congress and before the President of the United States, or your state legislature or governor. If you approve or disapprove, have enough belief in your sport to write your congressman. It doesn't have to be a long detailed letter—simply, "I'm for it" or "I'm against it."

Thirty-nine million fishermen. If 10 percent wrote, Congress would get 3.9 million letters. Keep a watchdog eye on those bills that have already been passed. The Wild Rivers Act seeks to preserve a few of our rivers in a "forever wild state." If there is a river in your territory being talked about for inclusion, or if you think a river you know should be included, tell your friends, go to work, support it, or suggest this new river to your congressman.

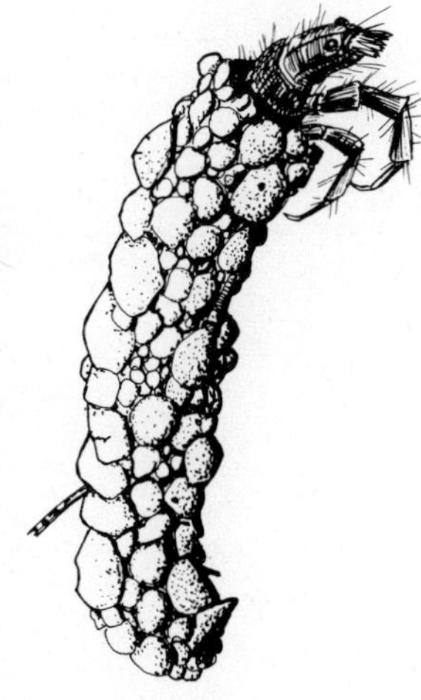

For some of our rivers the government has already allocated vast sums of money to clean up pollution. Certainly, this is a great step forward. But where there are millions of dollars involved, there are also bound to be chiselers. Be interested enough to see that at least on this one issue, there is no pork-barreling. How? Write your representative or senator when your state gets its appropriations. Ask how the proposed monies are to be spent, how much has already been spent. As a *citizen* fisherman, you will be surprised at the detailed information available to you just for the asking. If you don't like figures, get an accountant friend or a businessman to help you review them. Interest by a great number of people is the finest policeman in the world. And above all vote for your representatives on their record as conservationists.

In many communities, there are fishing clubs, some local,

some with national affiliations. Most of these clubs, or their local chapters, have an interest in some home water-conservation project and they do extremely valuable work. Within a few hundred miles of New York City, important conservation work has already been done on the Beaverkill, the Battenkill, the Amawalk, the Croton system, the Saddle River, and others. Without these groups, our flowing waters would be in a much sorrier state than they are now. If you are not a member, join. If you are a member and are helping in water projects—great. But if you are like the majority of club members in the clubs to which I belong, a few dedicated troopers do all the work and the balance give lip service.

How much better it would be, how much more would be accomplished if more members believed and worked at practical stream conservation. If you do not have the leisure or do not want to do physical work, there are many avenues open in club-sponsored stream-project work that as yet haven't truly been explored. Many times money is the necessary but lacking ingredient. An idea that I've been working on and fellow club members are developing with me is forcing gift participation by local businesses, big and small. It is almost useless to talk to the head of a concern and ask him to contribute part of his company's tax-free charity dollars to a stream project. His answer would have to be, "Why should I give money to help a few of your people who like fishing?" If, instead, there was well-prepared documentation proving that water conservation benefited the entire community, then the business executive might listen and contribute. The rewards for his company could be great local publicity and perhaps a permanent bronze plaque on a stream.

4

Some of my angling companions are not "joiners," but they love rivers and quality fishing. These individuals are alarmed by what is happening to our streams and will work for water conservation. What can an individual do by himself? Here are a few examples.

Part of a New York trout stream flows through lush dairy pastureland, and most of the farmer-owners are more than gracious about granting anglers permission to fish. Unlike some

other more fortunate meadow streams, the limestones, this stream's origins are in the mountains; snow melts and heavy rainfalls cause the waters to crest and fall rapidly. A few trees used to grow along the meadow banks, but gradually the current's erosion cut the soil from under their roots and they fell. Cattle pastured right into the stream, sliding into the water to drink whenever they felt thirsty—and the banks kept crumbling. The waters still held trout, but the stream was muddy a great deal of the time.

A friend of mine and I used to "trout" this section regularly. Gradually I found more romantic spots, but my buddy continued to spend most of his weekends casting for the large browns for which this section of stream was famous. Three years ago I joined him for a reunion. Where we always parked our car, there now was a new sign: "There is a $5.00 charge for fishing these waters, please see the owner." As I opened my mouth to complain, my friend smiled and said, "Before you holler, go pay your five bucks. I don't have to go with you. I run a charge account."

After being graciously received at the farmhouse, I returned with a piece of paper in my hand. This is a summary of what it said:

> The money you give us is spent strictly for stream improvement. Last year we took in $595.00.
>
> $437.65 was spent for fencing the stream.
> $196.43 for wire
> 137.15 for posts
> 104.07 for labor at $2.00 per hour
> 157.35 was spent for trees and planting
> ______
> $595.00
>
> This year with your contributions, we will plant more trees and put in some cribbing. I hope you enjoy better fishing.

While we were suiting up, my friend told me that once he had the idea it wasn't too hard to convince the farmer-owner. He showed him how his topsoil was washing away; how fencing a strip of land to protect the stream wouldn't take away much pasture. If he ever wanted to sell, a good protected trout stream would add many dollars to his price. Above all, my friend proved that improving his property wouldn't cost the farmer a cent. If any fisherman didn't want to pay five dollars a day for better fishing, he shouldn't be on his land anyway. And the best part of the whole deal was that the landowner had now convinced two of his neighbors to do the same thing.

When we got to the stream I saw at once that both banks were protected for twenty yards by a fence. The cattle had ingress and a crossing at two places only. In the protected strip, trees were beginning to grow and the roots of the alders and willows were already preventing wash. The stream was much less muddy—and the fishing was much better.

Another friend of mine is building a fishing cabin on the banks of a fairly remote stream. This is the third fishing lodge erected in an area where five years ago there wasn't even a jeep trail. This conservationist, at an extra cost, installed one of the new mineral-oil, self-contained, self-cleaning sewerage systems, so as not to risk contaminating the waters. No law coerced him to do this, but he had a basic concern for the environment.

Another man owns a small contracting business. At no cost to the state, for one or two days each year, he donates one of his small "cats" and his services as operator. Under state supervision, he repairs damage or works on other stream improvements. I never met this man, but I am grateful to him—and I've written him, as a concerned citizen, thanking him for his efforts.

And there is the example of Art Flick, who does battle constantly against the threatening forces of pollution, but takes time out every spring from his fishing to get more than two hundred willow shoots from the Conservation Department. These he plants himself along the banks of a public stream. Over the years, these have grown; the control they now exercise over flood damage is remarkable. The banks of the stream are contained, the water runs quick and cold, and the fishing is markedly improved. The Quill Gordons came back to this river this year, in good numbers, after Art had almost given up on their return. On a cold day in April, when it was too cold for them to get off the water he picked them up by the handful, warmed them with his breath, and carried them safely to streamside brush where they could rest and then mate.

There are many unsung heroes. The encouraging factor is that every month there are more and more concerned fishermen from all walks of life who are using their imagination, their strength, their monies, and their willpower to contribute.

More trout streams were ruined by bulldozers during the cleanup after Hurricane Agnes than by the floods that this sort of channelization purports to prevent. This sort of government response to natural disasters is both misguided and ecologically disastrous. It falls to the fisherman, and all other people who enjoy the out-of-doors, to provide their government agencies with another perspective.

For me, one of the best examples of individual action is a Chiracau Apache living in the White Mountains in Arizona. Now in his early seventies, he will still go with "sports" he has known or guided before on the Reservation to show them great trout, deer, or bear. Several years ago, he took me to the Black River. While I was joining my rod, he suddenly left me and waded well out into the river. Plunging his arm so deep that his face was also submerged, he retrieved an aluminum beer can. Without a word, he stamped it flat and put it in his saddle bags. From these same pockets he took one sheet from a small sheath of papers, unfolded it, and with a nail and a rock tacked it conspicuously to a nearby tree. His printed notice read: "Caught and picked up on this spot, one tin can. Are you the person who left it? For you or anybody else, cans and garbage do not make the river more beautiful."

I asked him where he had the signs made. He told me his wife was a schoolteacher and the school had a small press for its use. At their own expense, she printed several hundred for him. He used to have to put up a great many signs, but he thought they had done some good because he found people less careless now. Then he said something I remember well. "Most of the people we let on the Reservation are fine poeple, but some of them don't know better, or are so selfish they don't give a damn. Someday, everybody will have to learn to care. Rivers are so beautiful."

And this, I think, is what is happening in water conservation today. More and more people are learning to care. This is especially true about fishermen. Until the moral ethic is more widely accepted, more anglers must be willing to be their brothers' keepers. Perhaps they will come home from a day on the stream with their creels bulging with plastic containers or flattened cans instead of too many fish. Perhaps they can learn to watch water with more care, and spot the trouble early. The eddy pool is washing too much? "If I can get permission from the Conservation Department, I have a friend who could haul a load of stone we could put in the right place." A small mill is dumping stuff in the river again? "I'm going to put in a call and get the right people to stop it."

Thirty-nine million fishermen. We can hold our flowing waters where they are today and, if all of us work, improve them.

We can become "true" and loyal "companions" of the waters we fish—and *must*—or fishermen will be an endangered species.

Wild fish leaping the falls in a wild river: what we all love to see.

2

The Living Stream

WILLIAM FLICK

Everything is hitched to everything else. If man doesn't act as an integral part of a harmonious whole, he'll bring unbalance and beget ultimate loss and poverty.
JOHN MUIR

Mountains are majestic. From their peaks the world lies below in serene splendor, but the deathly quiet indicates little of the life that throbs within their reaches. It is in the stream that runs from the mountain to the sea that life appears in a multitude of forms and is most obvious in a quiet and subtle way. As we sit by a wild mountain stream, watching it flow over the stones and among the trees, we realize the area teems with life. A kingfisher flies by in search of a small fish, and trout rise to hatching mayflies. Among the stones, larval stages of numerous insect species are grazing on minute plants, or straining food from the water. Luckily this particular stream and its watershed has been undisturbed by man's activities. In this respect it is, unfortunately, unique.

The life within the stream is interrelated and complex, with some members of the aquatic community needing a series of habitats to carry out their life cycle. A large trout, for instance, may live in the depth of a shaded pool but move into another area for feeding, and still another for spawning. The prevalence of trout, or other stream animals, depends upon available food within the stream system. Loss of a segment of the stream community, through habitat deterioration, food supply shortage, or destruction of areas for reproduction, will affect not only that particular animal but the others as well. This fact is extremely important and cannot be stressed too strongly. Let's look closely at this wild stream and determine what factors influence its course, its flow, and some of the basic ecology of the life within it and thus be able to understand better the fragile nature of this segment of our environment—what keeps it healthy, and how man's activities can affect not only the stream but the life around it.

PHYSICAL ASPECTS

Our stream is born at a small spring up on the mountain. This water is joined by that from surface runoff, resulting from rain or melting snow, plus water from other springs along the stream course. As the water flows down the mountain, it takes the path of least resistance and where there are rock formations or bank sediments that are tightly compacted the stream will flow around them. If the gradient is steep, the stream velocity will be high; erosion of the bottom and bank material will occur in direct proportion to the force of the water and the size, compactness, and erodibility of the material against which it strikes. The gradient over which the stream passes varies and as the water flows from steep to more gradual slopes, or over soft bottom sediments, pools are scoured out and the eroded material deposited at the lower end. If a stream changes direction, a pool will form at the bend because of a spiral-type circulation that scours the bottom to a degree greater than that caused by velocity alone. In such a situation, if the stream is bending to the left, the greater velocity will be on the right bank, as will the deeper part of the pool. With an area of lesser velocity on the opposite side, sand and gravel will settle, forming bars; the coarser material will be deposited first, followed by finer material being deposited inward and upward on the bar where currents are slower.[1]

As the currents strike the opposite bank, the direction of flow is again changed and the sequence reoccurs, resulting in a meandering course. It is only in the areas of very steep gradient, where the velocity is sufficient to cut through the banks, that the stream maintains a straight course. Thus, in the lower reaches of the stream where the gradient and velocity are less, the stream develops a distinct meandering course. In all but the most precipitous sections, this normally creates a series of alternating pools and riffles. Under natural conditions, one seldom sees a perfectly straight stream, nor, except in unusually steep gradient or level terrain, one of continuous rapids or pools. The alternate pool-riffle relationship has distinct advantages in that it forms a diversification of environments, which is essential to a high population of aquatic life. The meander course is also desirable because it slows the stream's velocity and thus decreases erosion. Straightening or channelizing a stream is very much to the detriment of its aquatic life.

In a wild stream the banks have stabilized over the years and are held in place by the roots of grasses, shrubs, and trees. With the stream course stabilized, little further bottom scouring occurs and the stream bed remains essentially the same from year to year. The abundant grass and forest cover throughout the watershed soak up and hold back rainfall and melting snow, and there is little erosion to bring silt into the stream system. The water therefore is clear, and water temperatures remain cool due to abundant shade and numerous springs that slowly release groundwater from prior precipitation. In this stable environment a multitude of life forms have evolved, with adaptations to survive in the many different habitats or niches available.

THE AQUATIC COMMUNITY

Water, the medium in which life originally began, is a unique substance. It harbors plants and animals that are very simple in structure and others that are very complex. The various forms have a direct relationship with one another, a fact not often realized by the average fisherman or hiker who travels along the stream. They see the mayflies and other larger insects

Nature's laws will not be denied. Although this channelization project completely disrupted the biological life systems of Cove Creek in North Carolina, sandbars (the white areas in the photo) are already developing from the erosion of banks. *North Carolina Wildlife Resources Commission*

and trout that feed on them. A mink scurrying along the water's edge searching for crayfish is noted with interest, but the lesser forms of life are not often appreciated for the vital role they play in the stream community. To understand the life in our stream more fully and what keeps it healthy, let's look through the eyes of a biologist and see what transpires beneath the surface. Perhaps we can learn some of its secrets.

Nutrients

Water is a solvent and dissolves more substances, both gaseous and solid, than any other liquid. This characteristic is of vital importance as it is the dissolved nutrients and gases, together with radiant energy from sunlight, that governs the abundance of life in the stream.

Rainwater contains significant amounts of nitrogen, one of the elements important in plant growth. In addition, it picks up carbon dioxide from the air and topsoil, making it acidic, because of the formation of carbonic acid. This enables water to dissolve more substances than "pure" water, which actually does not exist in nature. One such mineral that is dissolved is calcium carbonate from limestone or calcareous soil deposits. The resulting calcium bicarbonate is very important for its buffering effect. Along with the bicarbonate salts the water also picks up the other elements necessary for plant growth, such as carbon, hydrogen, oxygen, sulfur, phosphorus, magnesium, potassium, and iron, plus certain trace elements. The concentration of these minerals will vary from one part of the country to another.

Streams found in areas of ancient rock formations usually have been so severely weathered that most of the soluble minerals have been leached away. These waters have low concentrations of all the major ions, and the water is soft due to the low concentrations of calcium and magnesium. They are poorly buffered because of the low carbonate content and hence are acidic due to small quantities of organic and inorganic acids derived from the soil and vegetation in the watershed. In contrast, water flowing through limestone areas is high in calcium carbonate and the other minerals necessary for plant growth. These waters are normally slightly alkaline.

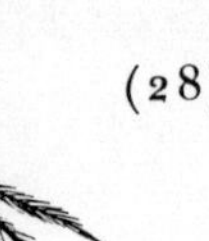

The term pH is used to connote the relative alkalinity or acidity and is given a range of 0 to 14. The point of neutrality (equal amounts of H+ and OH—) is pH 7. Above this there is an excess of hydroxyl ions and the solution is alkaline, while below this there is an excess of hydrogen ions and the solution is acid. Waters that are excessively acid (below 5.0) or alkaline (above 8.5) are not conducive to good growth and survival of most salmonid species and are approaching the tolerance limits for many plants and invertebrates. As mentioned earlier, waters that are slightly alkaline are usually rich in dissolved nutrients, and pH is thus indicative of relative productivity.[2]

Food Web

The life in our stream starts with the basic nutrients, plus radiant energy, to produce plant tissue, which is in turn utilized by other members of the aquatic community. This flow of energy and nutrients through a series of organisms is referred to as a food chain. In the aquatic environment this transfer becomes quite complex as it goes, for example, from plants to insects to fish and eventually back to the basic nutrients. Because of its complexity (see Figure 1), a more appropriate and descriptive term is the food web.

The various segments of the food web are classified by the energy source and nutrients utilized. Each segment is referred to as a trophic level, the first being the primary producers that utilize inorganic nutrients and radiant energy to produce plant tissue.

We will follow the flow of energy through the food web or chain until we reach man and the mammals and birds that utilize the fish.

The inhabitants of the fast-water area are quite different from those found in pools. This particular section of stream is near the headwaters in a northern latitude, and harbors cold-water species such as brook trout and slimy sculpin. If we were to move to warmer waters, the trophic levels would remain the same but the species involved might be different. For example, temperatures in the 70s (degrees Fahrenheit) for long periods of time are not favorable for trout survival, and bass or pike might replace this species. The food web would be basically the

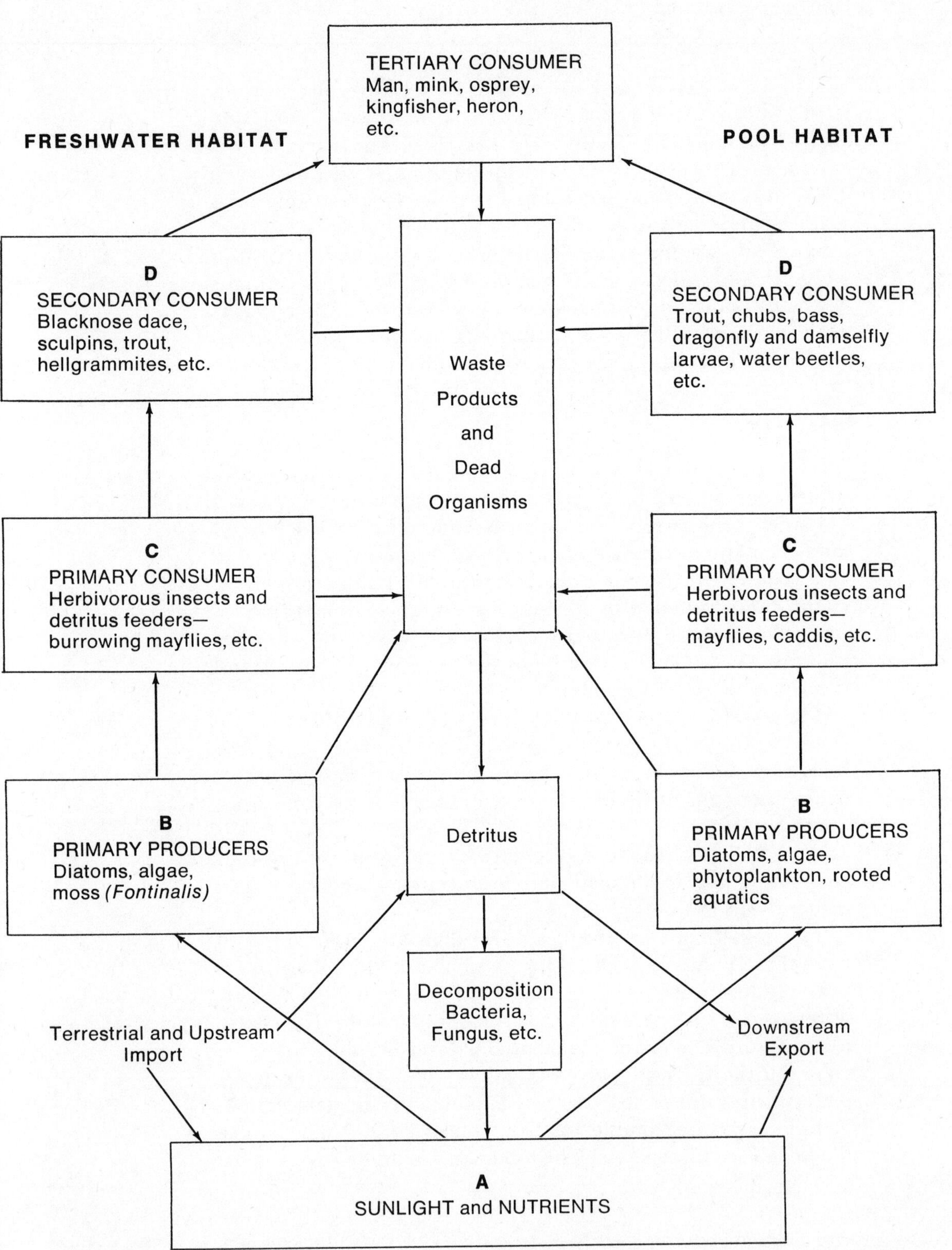

TERTIARY CONSUMER
Man, mink, osprey, kingfisher, heron, etc.
FRESHWATER HABITAT
POOL HABITAT
D
SECONDARY CONSUMER
Blacknose dace, sculpins, trout, hellgrammites, etc.
Waste Products and Dead Organisms
D
SECONDARY CONSUMER
Trout, chubs, bass, dragonfly and damselfly larvae, water beetles, etc.
C
PRIMARY CONSUMER
Herbivorous insects and detritus feeders—burrowing mayflies, etc.
C
PRIMARY CONSUMER
Herbivorous insects and detritus feeders—mayflies, caddis, etc.
B
PRIMARY PRODUCERS
Diatoms, algae, moss (Fontinalis)
Detritus
B
PRIMARY PRODUCERS
Diatoms, algae, phytoplankton, rooted aquatics
Decomposition
Bacteria, Fungus, etc.
Terrestrial and Upstream Import
Downstream Export
A
SUNLIGHT and NUTRIENTS

THE FOOD WEB

THE FOOD WEB

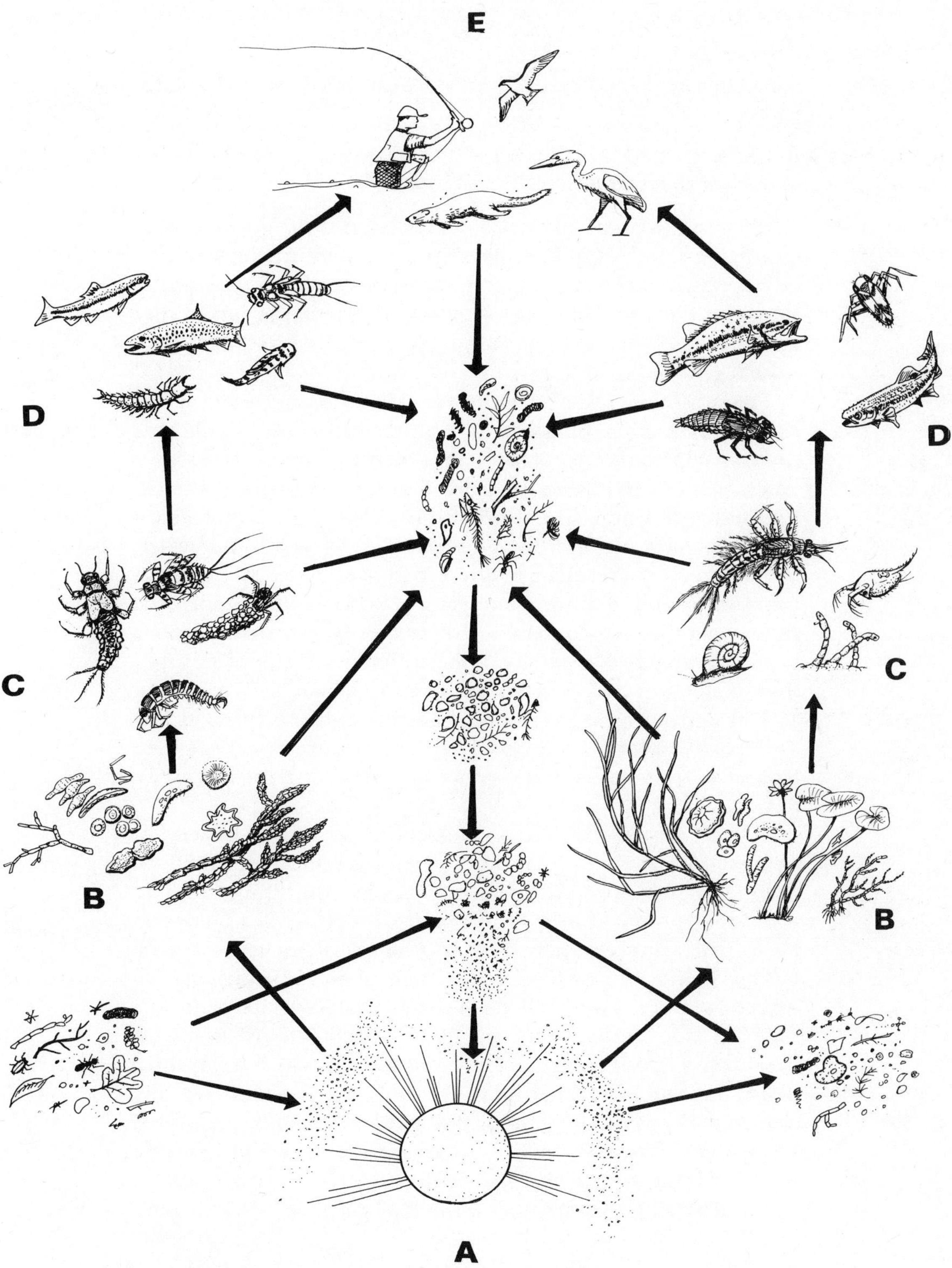

An example of the food web in a stream showing radiant energy (A) from the sun along with the import of nutrients from the food web and upstream areas, along with the loss of nutrients through downstream export. Primary producers are shown in (B); primary consumers (C); secondary consumers (D); and tertiary consumers (E).

same, however, and could just as easily be destroyed or disrupted.

Fast-water Community

The fast-water or riffle section of a stream is the greatest food-producing area, as the flowing water transports nutrients to and carries waste products away from the many aquatic organisms at a quicker rate. The moving water is also very important in that it replenishes the oxygen supply that is vital to plant and animal growth.[3]

The substrate of the fast-water area is also an important factor in the production of plant and animal life. Flat, shale-type stone, particularly that found in limestone areas, provides a much greater surface area for growth of minute plants than does the sand or fine gravel that is found in slower water. When such flat stones have much of the undersurface exposed to the water, mayflies, which are of such interest to the angler, abound. In contrast, round granite boulders or bedrock have only the surface exposed and harbor fewer mobile insects of the mayfly type. Our stream happens to be in a limestone area with a pH slightly above 7 and is much richer in life than a stream in an area of igneous rock where the pH is apt to be on the acid side (pH below 7) and where dissolved minerals are low. Let's then look closely at our bottom stones and see what organisms we have.

Primary Producers. Throughout the stream a variety of plants occur, with each section harboring some types that are unique to that area. All green plants have one thing in common—they produce tissue by a complex process called photosynthesis. This process utilizes energy from sunlight to synthesize fats, proteins, and carbohydrates from dissolved nutrients absorbed from the water or bottom sediments. As we look into the water in front of us we do not find any rooted aquatic plants. The bottom sediments and the current are such that these plants cannot exist here, and the only types that have any chance for survival are those that are able to cling to the stones. If we were to scrape some off one of the stones and examine them under a microscope, we would discover there are several forms of algae but in particular the type referred to as diatoms.

RIFFLE OR FAST-WATER AREA

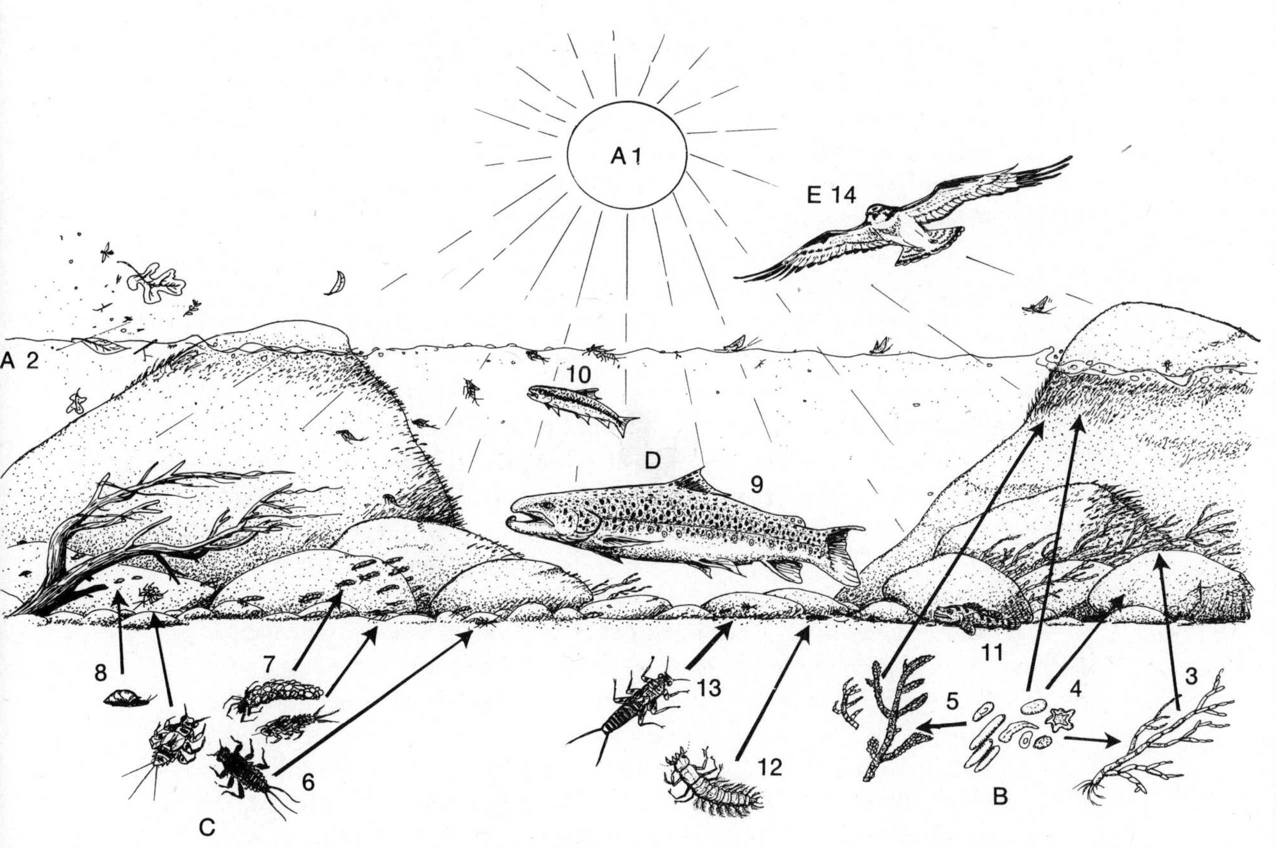

Life in a riffle or fast-water area of a stream compared with that of a slow-water area. The letters indicate the trophic level of the various groups, as shown in figure 2. Individuals of the various trophic levels are grouped for illustrative purposes, but in nature would be intermixed. Fast water: A (1) sunlight and radiant energy; (2) nutrients; B (3) algae *Cladophora;* (4) diatoms; (5) water moss *Fontinalis;* C (6) mayfly nymphs; (7) caddis fly; (8) snail; D (9) brown trout; (10) red-belly dace; (11) sculpin; (12) hellgrammite; (13) stone-fly nymph; E (14) osprey.

These single-celled forms of algae are characterized by a cell wall of silica which is covered with a jellylike substance.[4] Although they are among the most abundant of all living things in all waters of the earth, the term *diatom* may be new to many fishermen. It is they, along with some blue-green and green algae, that make the stones so slippery in many of our streams. Their role is not to impede the angler from reaching the opposite shore but rather to provide a source of food for the herbivores that live in this environment.

On our stone we may also find a small greenish moss (genus *Fontinalis*). The abundance of these producers is determined by the quantity of inorganic compounds dissolved in the water, water temperature, and pH.

Primary Consumers. The herbivores (plant eaters) and others at the consumer level use organic nutrients and chemical energy to produce tissue, in contrast to the preceding trophic level which utilized basic nutrients and radiant energy. The herbivores living among stones in the rapids are especially adapted for this particular environment. Their bodies are generally flattened to provide less resistance to the current, and they often have hooks or suckers to enable them to grip smooth surfaces. The caddis and larvae of the blackfly are not as flat as the mayflies and stone flies, but are especially adapted to withstand strong currents. For example, the blackfly larvae not only has a sucker at the posterior end of its body but it also attaches itself to bottom material by means of a silken thread. If dislodged, the larvae is kept from being washed downstream by this silken "safety rope" and can thus work itself back to another favorable location for attachment. Some caddis flies use hooks to grasp irregularities on the stones, while others construct cases of sand or small gravel to protect themselves from the current. Still others live in funnel-shaped nets that they spin about themselves and cement to stones. These nets not only prevent them from being washed away but also trap food that is brought down by the currents.

If we look at both sides of a large flat stone we will be awed by the multitude of community life that exists in this small area. Of particular interest to the angler will be the mayflies. We will find the streamline form, represented by *Baetis,* and the flattened types such as *Iron* and *Stenonema.* We will also find blackfly larvae, caddis, snails, stone flies, scuds, water pennies, and others. The mayflies are of particular importance as food for the members of the next trophic level as they are mobile and in their movements become vulnerable to the various predators. In addition, many of the invertebrates we have just discussed have a winged stage, and during the period that they move from the bottom to the surface to hatch they become easy prey for the trout and minnows, as well as for the birds which live along the stream banks. It must be remembered that many of the primary consumers are small and fragile, but extremely important members of the aquatic community and do not have the adaptability or maneuverability of the larger predatory animals at the next trophic level.

Secondary Consumers. The major secondary consumers are the fish species and the predaceous insects that live in the riffles and prey on the mayflies and other invertebrates. Predaceous invertebrates are not generally as common in the fast water as in the pool section—the two major species being the hellgrammites and the stone flies. Admittedly, some of the caddis in the primary consumer level occasionally eat some animals, but they are not considered as predaceous as the hellgrammites, stone flies, or the fish. The sculpin and blacknose dace are the two minnow species that usually inhabit the riffle areas and feed on the smaller invertebrates, along with some of the primary producers. In this same environment we will also have brook trout, or salmonid species that feed primarily on the invertebrates of the preceding trophic level.

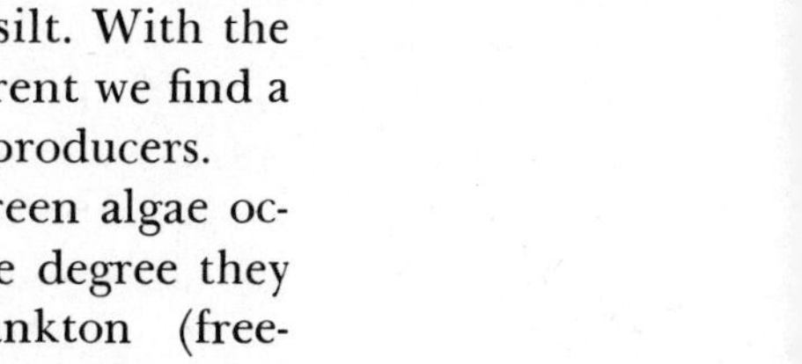

Pool Community

Let's move downstream to a long quiet pool. Here the lack of current allows suspended material to settle out and the bottom composition contains more fine gravel, sand, and silt. With the change in bottom composition and the lack of current we find a change in the abundance and types of the primary producers.

Primary Producers. Diatoms, blue-green and green algae occur in the quiet waters of the pool but not to the degree they were found in the riffles and rapids. Phytoplankton (free-floating microscopic plants) are not found in the riffle area, except drifting through, but may occur in quiet pools, particu-

SLOW-WATER AREA

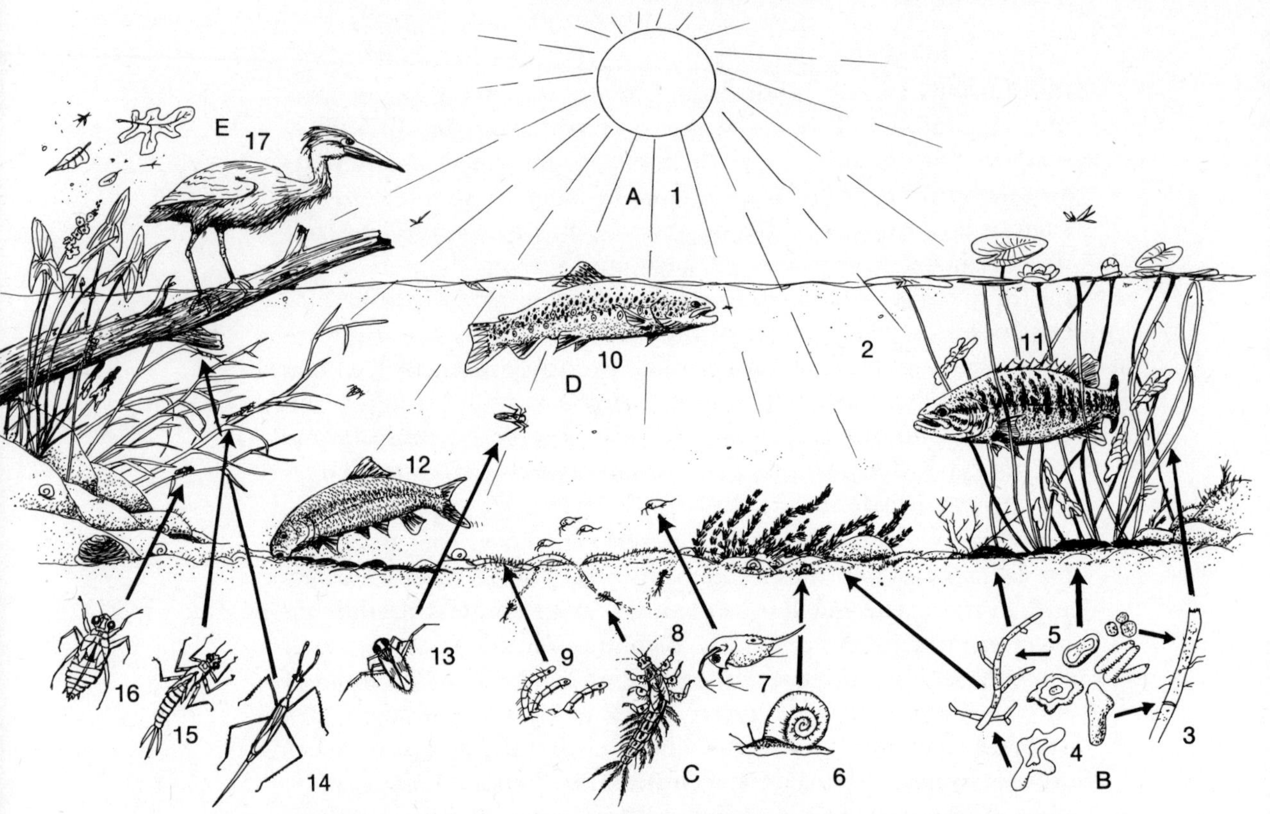

Slow water: A (1) sunlight and radiant energy; (2) nutrients; B (3) spatterdock Nuphar; (4) diatoms; (5) algae *Cladophora;* C (6) snail; (7) zooplankton *Daphnia;* (8) burrowing mayfly nymph *Hexagenia;* (9) bloodworm; D (10) brown trout; (11) bass; (12) sucker; (13) water boatman; (14) water scorpion; (15) damselfly nymph; (16) dragonfly nymph; E (17) heron. (The illustrations are not to scale. This and Plate 2 are adapted by permission from Dr. Smith's *Ecology and Field Biology,* Harper & Row, Publishers, Inc.)

larly if there are backwater areas, or ponds and lakes that drain into the system.

If the pools are long and not subject to strong currents, rooted plants may appear, giving a pond-type environment. These rooted aquatics are usually covered with diatoms and other algae that are food for the primary consumers. The aquatic plants that penetrate the surface are not only the habitat for the insects themselves but also may be areas for egg laying by some dragonflies, damselflies, and others. Their greatest importance, however, is as the home of many of the larger insects.

Primary Consumers. The silt and decaying organic matter in the pool harbors different organisms from those found in the stones of the fast water.[5] Instead of insects adapted for clinging to the stones, we find those that burrow in the silt or build tubes of organic material. The most beautiful of our mayflies, *Ephemera* (Green Drake) and *Hexagenia* ("Michigan Caddis"), have flanged lobelike legs permitting them to burrow into the bottom. In the pool habitat, caddis may be numerous but they are the species that build their cases from plant material rather than from stone as was the case in the faster water. Larvae of midges (bloodworms) are not often seen by the casual fisherman but they may be abundant in the bottom sediments in their small tubes.

In contrast to rapids, which have mainly herbivores that graze on living plants, many insects in the pool habitat are detritus eaters, feeding on dead organic material. The detritus in the pool may have originated there, drifted down from the fast water, or resulted from leaves or other material blown in from the terrestrial environment. Thus a pool does not necessarily produce all its own food.

Although zooplankton (free-swimming microscopic animals) are not as common in streams as in ponds and lakes, they may be present and are extremely interesting, particularly when examined under a microscope. They come in many bizarre forms and shapes, some with their young in tiny pouches on their backs. They are excellent forage for small fish and other small predators in the secondary consumer level, and feed on phytoplankton, bacteria, and detritus.

Secondary Consumers. More predaceous species live in the pool area than in the riffles, as here their larger and bulkier bodies are not affected by currents that would restrict their movement in catching prey. In this group are included crayfish, dragon- and damselflies, as well as several species of water beetles. Although they may not be as abundant as the mayflies, their large bodies make them important food items for larger fish.

The fish population in the pools also differs somewhat from the riffle and rapids area. Here you will find members of the sucker family, which feed on both the primary producers and primary consumers. They are well adapted, with their sucking mouth parts, to utilizing food on the bottom. This species often makes up a large percentage of the biomass in the stream. Minnow species found in pools, such as the creek chub and the fallfish, reach a larger size than minnow species found in faster water. Larger trout are often found in pools, where they can seek cover in the deeper water and feed on the larger food items. They are not restricted to larger insects, however, as they may feed actively on the smallest mayflies or midges among the primary consumers. One often wonders how they can obtain sufficient food value from these minutae to compensate for the energy expended in catching them. Larger trout are not restricted exclusively to pools, for they may move into riffles to feed and, to spawn, must find a riffle or area where water percolates through the gravel.

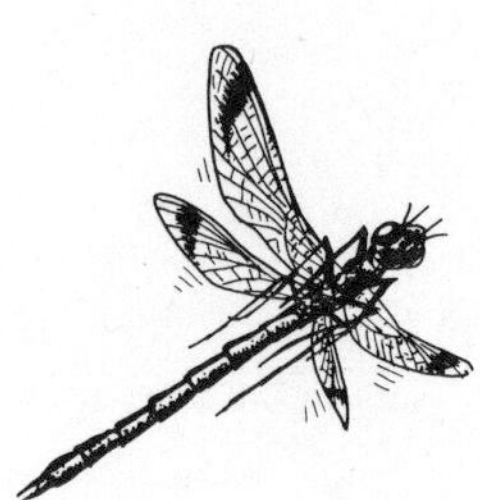

The trout, or warmwater game fish found in this trophic level, whether they are in the pool or fast-water habitat are of particular interest, as they have provided sport, as well as food, to mankind for hundreds of years. In a wild river that has not been degraded by man, they show an amazing ability to sustain themselves.

Tertiary Consumers. At the end of the food chain we have predaceous birds and mammals that feed on fish, many of which are interesting members of our stream community. Our stream would hardly be complete without the rattle of a kingfisher calling, or the quiet movements of a mink as it looks for a crayfish or a small trout or minnow. A few years ago we might have observed a bald eagle as it attempted to take a fish away from an osprey, but unfortunately bald eagles are now rare in most areas. Man himself falls into this trophic level, as he uses some of the fish species as food. He is the one segment of the food chain that has increased in numbers, but not necessarily to the advantage of the others. The tertiary consumers are the most maneuverable of the group in the food chain and they are as likely to be found in the fast-water area as in the pool, or moving from one to the other and back again.

Decomposition and Mineralization

A very important process in the flow of energy within the food web is the breakdown of organic material by decomposers, so that it is returned as basic inorganic nutrients to be recycled again by the primary producers. Waste products from the various organisms, as well as their bodies following death, are decomposed by bacteria, fungi, and other decomposer organisms which sequentially break down carbohydrates, proteins, and fats into simple inorganic products such as carbon dioxide, minerals, and salts. Although nutrients are returned to the stream system through this process, some are lost as they are washed, or exported, downstream. Likewise, new ones are blown in from the terrestrial environment or washed down from upstream areas. The aquatic environment thus remains healthy as long as each trophic level is maintained and functions normally.

Biological and Chemical Relationships

In the previous discussions we did not detail the chemical and physical conditions that influence the well-being of the various organisms. Obviously such factors are important if we are to appreciate fully the delicate nature of the various organisms and how they may be affected by environmental changes.

In discussing nutrients, the importance of the buffering effect of bicarbonate salts was mentioned. If streams that are naturally acidic, and thus low in buffering agents, receive pollutants that are acidic, or minerals such as zinc or copper, toxic conditions may develop from even very small concentrations. These waters are already low in productivity and can hardly stand further reductions in plant or animal life if we hope to have a healthy food web.

At the bottom of the food chain we saw the importance of plants as a source of food for the herbivores. In a sense, our stream is a garden of small plants, and as with our home garden a large number of elements are necessary for good plant growth. If even one element is lacking, or in an insoluble form, the entire food chain is affected.

Temperature and sunlight are also important as plants grow faster at higher temperatures and sunlight is necessary for photosynthesis. For this reason, headwater sections of streams, which are usually shaded and cool, are not as productive as areas farther downstream where the stream is less shaded, assuming of course the nutrient base is similar in both areas.

Unpolluted water, like unpolluted soil, is also important. We

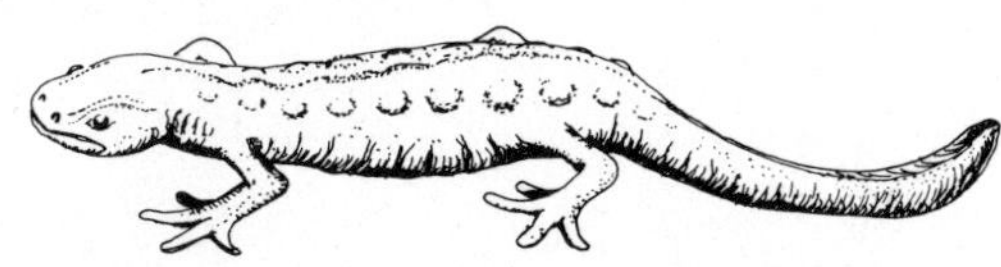

Every stream—large and small—functions as a drain for all the soluble chemicals and materials in its drainage. What is not absorbed by the vegetation will seep through the water table into the stream's system. It is a basic law of nature that varies only in scale, and when the scale becomes large enough even man notices its effects. This last untamed stretch of the Missouri now lies under the threat of channelization. *Nebraska Game and Parks Commission*

would not think of dumping pollutants in our home garden but often the stream is considered an ideal spot for disposing of waste material. *Any pollutants that adversely alter pH, temperature, light, oxygen, or availability of basic nutrients will affect the abundance of plant life.* The plants may be so minute that they can barely be seen, and their destruction may go unnoticed by man, but not by the animals that depend on them for food.

The presence of dissolved oxygen is of prime importance to most members of the aquatic community. Fortunately, water is aerated as it passes over riffle sections, and in a wild stream oxygen values will be near saturation, which is approximately 9.0 ppm (parts per million) at 70° F and increases to 14.6 ppm at 32° F. It is only when a stream has been subject to degradation that oxygen values may become critical. If dissolved oxygen is reduced, some of the riffle insects may die unnoticed before the relatively tolerant trout, which can survive at approximately 4.0 ppm (parts per million) under cool-water temperatures.

A stream low in nutrients will be low in aquatic plants and animals, but it is also possible to have a stream too rich in nutrients. The latter does not normally occur under natural conditions, but where sewage or other sources of nitrates and phosphates reach the stream, as the result of man's activities, the stream may be degraded. For example, in a slow-moving stream that receives domestic sewage, or receives water from overly enriched lakes, dense populations of phytoplankton may cause the water to turn green or brown, or heavy growths of filamentous algae may develop. Unusually heavy algae blooms are not a sign of a healthy river; they may seriously reduce dissolved oxygen during periods of darkness, or, occasionally following a large increase in algae abundance, heavy mortality of these same plants may suddenly occur resulting in low oxygen values because of oxygen demands during their decomposition. As mentioned earlier, the inhabitants of the fast-water areas cannot tolerate low oxygen and a riffle below a pool with an intense bloom can be affected by reduction in dissolved oxygen, as may the fish species throughout the stream if conditions become extreme.

In a wild stream the fish species have an amazing ability to sustain themselves even in the face of moderate exploitation. The reproductive potential is high and a one-pound female brook trout, for example, may deposit a thousand eggs in its redd. In a stabilized stream bed, water percolates through the gravel throughout the winter and in the spring over 90 percent of the eggs may hatch. This aspect of the life cycle is extremely critical as the eggs or newly hatched fry are unable to move to a new location if conditions become unfavorable. In a healthy stream there will be little siltation or scouring of the riffle areas where spawning occurred and the newly hatched fry can develop normally; when they emerge from their gravel nest they find abundant food and environmental conditions good for survival. It is when we get into a stream that has been degraded that we may find this stage of the life cycle most limiting; then the fish supply must be supplemented through stocking.

There probably should be one final category in our story although you will not find it in most texts that discuss food chains or food webs. It affects the survival and health of the entire aquatic community and may be the most important category of all. This final category is the primary destroyer. Man is the sole member. Without man's deleterious influence, the environment and food web would function normally year after year with only minor changes due to occasional floods or droughts, and even these would be but temporary wounds. But if one link of the chain is broken by man the entire aquatic community is affected.

PRIMARY DESTROYER

We have been discussing the life in a stream in a wild setting, unaffected by activities of man. What are some of the problems that develop as the land is cleared and developed, supposedly for the benefit of mankind?

The stability of the environment is extremely important in maintaining a healthy food web as in many aquatic species the life cycle is very involved, going through egg deposition and one or more larval stages before the adult form is reached. This may take from several weeks to several years, depending on the species. In these various stages the requirements for survival are very rigid, and even slight changes in the environment can be catastrophic. In the headwaters of most streams there is a timber resource that can be harvested. If this forest cover is removed, a chain of events occurs throughout the stream. The first bit of erosion brings in new sediments. As these settle out, a change in current patterns develops, new bars are formed, and scouring occurs at new locations. With each shifting of bottom sediments, changes in habitat occur. Where we once had a quiet pool, we now have a sandbar or possibly a riffle. Obviously, animals and plants that can not adjust to changes in environment must perish. Species that do survive find their food supply reduced or eliminated and spawning areas covered with silt or scoured away.

A simple logging operation? Well, yes, a simple logging operation that will result in erosion and eventually an alteration of the oxygen, sunlight, and ultimately the whole biology of the stream. That is the context within which we must put all our acts: all things are connected. *California Department of Fish and Game*

With removal of the water-holding qualities of a forest cover, or grass cover in the case of poor farming practices, water levels are subject to wide fluctuations. The primary producers (algae, diatoms, and so forth) are sedentary, and drops in water level may leave them exposed to the air to die, or, in contrast, flash floods will tear them from their positions on the stones. Changes in the environment may affect the higher members of the food chain as well. Decreases in groundwater and removal of forest cover allow water temperatures to increase and the fish species association may change. Where at one time trout appeared downstream for many miles, they may become restricted to the headwaters or close to springs.

Man's interest in utilizing natural resources often leads to the mining of various minerals. These activities may result in silting of the stream beds and smothering of the less mobile plants and animals; or toxic products may reach the stream system. A mining operation may be designed to prevent pollution, but roads leading to and from the mines are often built with gravel from the mining operation. If the mineral extracted happens to be zinc, copper, or some other toxic material, runoff over the road system may pick up these minerals and carry them to the stream with deleterious effects. If concentrations are light they may not actually kill fish but they may affect the organisms at the base of the food chain and result in a decrease in the productivity of the stream. In coal-mining areas, acid water from the mines may reach the stream system and lower the pH to the point where not only the animals and plants at the base of the food chain can not survive but, of equal importance, even the bacteria which break down the organic material in the stream are drastically reduced. Most streams under natural conditions will have a pH of between 5.5 and 8.0. In some mining areas of Pennsylvania and the Virginias, pHs below 5 are common and even the most acid-tolerant fish may be absent from such waters.

At times we take on the role of benefactor rather than destroyer, but often we do not look at the overall ecological effect of our actions. A good example is the use of insecticides to save our woodlands. In broad areas of Canada, DDT was used to destroy the spruce budworm. In the process this chlorinated hydrocarbon, which takes years to break down, entered the stream system and killed many of the aquatic insects that were important as food to a valuable salmon resource. In some instances the concentrations were so high that the fish themselves were killed. Even the tertiary consumer in the food chain may be affected. Fish-eating birds have been found to have accumulated high concentrations of DDT from their food source. This insecticide is blamed for the abnormally thin shells on the eggs from these birds and reproduction has been so low in recent years that in many areas ospreys and bald eagles have become rare.

The problem of dumping sewage in the stream systems is only too well known. Organic material in a stream system is immediately attacked by bacteria that use enormous amounts of dissolved oxygen in the process of decomposition. As mentioned earlier, our fast-water inhabitants live in an area where oxygen is normally near saturation. Decreases in dissolved oxygen, which may not kill fish, may destroy their food supply. A fast-flowing stream constantly picks up oxygen from the air and will purify itself quickly but in the lower reaches, where the pools are long and slow, this may not be possible. The normal ecology of the stream may therefore be completely changed by the oxygen decrease. Mayflies, for example, which need abundant oxygen, may be replaced by midge larvae, which can survive without oxygen. Aerobic bacteria are replaced by anaerobic forms and normal stream life is destroyed. Even in rural areas similar conditions can develop. Improperly placed manure piles can pollute a small stream just as seriously as domestic sewage.

Poor farming practices may result in soil erosion and stream siltation. It must be remembered that the primary energy source of all food webs is sunlight. If water transparency is reduced plant growth is inhibited and stream productivity is decreased. In addition, the material washed into the stream will settle out, smothering stream life.

The magnitude of man's ability to fill the role as primary destroyer knows no end. We can repair some of our mistakes, but we must understand the interrelationship between the many forms of life, or by correcting one wrong we are likely to create another. Our stream is truly teeming with life, but only as long as we keep it healthy and to do this we must understand its ecology.

A free-flowing natural stream speaks for itself. Must such beauty die? *Michigan Department of Conservation*

SIGNS OF STREAM STRESS, CAUSE AND CURE

How can you recognize signs of stream stress and what can you do to keep our streams healthy? Some situations may need the trained eye of an aquatic biologist to diagnose, but others will be obvious to the most casual observer. In all cases the advice of a trained biologist is recommended before any corrective measures are taken. Here are a few examples of unhealthy conditions that can be watched for and sometimes corrected:

INDICATOR	CAUSE (*Physical*)	CURE
Shifting stream bed Shifting gravel beds Eroding stream banks Widening stream bed Filling in of pools with gravel Siltation	Loss of bank cover by cutting trees and willows Destruction of bank cover by livestock Disturbance of stream by bulldozing Landfill operations Road and bridge construction	Planting bank cover and fencing Stream-improvement structures Stream-protection legislation
Increase in water temperatures beyond the norm expected in midsummer	Dams within the stream system (beaver dams included) Low water table Deforestation	Removing dams, or, in large impoundments, bottom drawoff Piping springwater to favorable locations or cleaning out springs Reforestation
Muddy water	Erosion of soils or disturbance of streams by bulldozing, etc.	Legislation
Absence of small trout or other game fish	Loss of spawning beds through siltation or dams	Stocking, stream stabilization, and/or fish ladders

	(Chemical)	
Unusually low pH levels Fish kills (more than an occasional fish) Absence of insect life in riffles (particularly mayflies) Abnormal water color	Industrial pollution Mining	Legislation and law enforcement
	(Organic)	
Heavy algae blooms (pea-soup color to water) Oxygen values considerably below saturation	Sewage, either domestic or from livestock or poultry	Legislation and law enforcement Citizen action

Notes

1. James C. Knox. "How Streams Are Shaped," *Trout* magazine, Winter 1972.
2. Carl L. Schofield. "Water Chemistry and Lake Productivity," *The Conservationist,* April–May 1970.
3. See Robert L. Smith's *Ecology and Field Biology,* New York: Harper & Row Publishers, Inc., 1966.
4. James G. Needham and J. T. Lloyd. *The Life of Inland Waters.* Ithaca, N.Y.: Comstock Publishing Co., 1937.
5. H. B. N. Hynes. *The Ecology of Running Waters.* Toronto: University of Toronto Press, 1972.

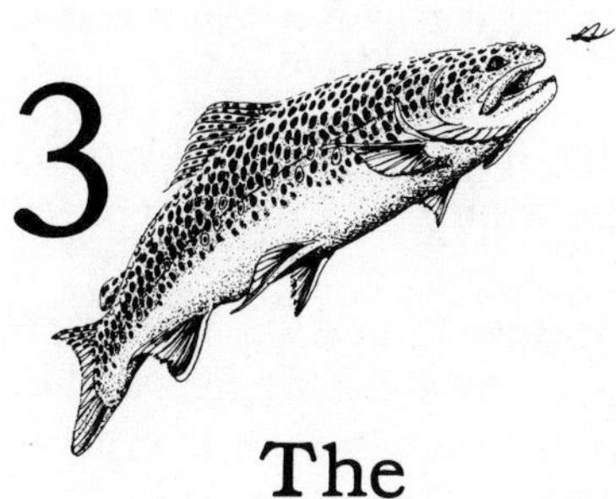

3 The Stream Killers

BEN EAST

If Man continues to pollute the earth at the present geometrically increasing rate, the time may come when cosmic nature will find it necessary to recycle the planet.
The Washington Daily News

The free-flowing, undefiled streams remaining in this country today are no more than a remnant of our original heritage of running water; the rest we have killed outright or altered irretrievably.

The list of what things destroy streams and rivers is so long and the consequences of some are so dire that it is no easy task to name those that pose the gravest threat. But pollution has to stand at or near the top.

Water pollution divides into two broad classes. The first, which has its source at a specific location, is called "point-source"; it includes such offenders as municipal wastes from sanitary sewers or inadequate sewage treatment plants, and industrial outfall. The second does not originate in any one place and so is known as "nonpoint-source pollution." Agricultural and urban runoff are good examples.

Of the two, point-source pollution is far easier for the average person to see and identify and for the angler-conservationist to do something about. Any fisherman who sees raw sewage floating in the water of his favorite stream, or watches oil or other industrial waste pouring in a murky stream from the mouth of a sewer pipe along the bank, knows what he is seeing, can back his complaint with proof, and is in a position to go to the proper authorities and demand a remedy.

On the other hand, destructive wastes entering a stream in the water of a drainage ditch, picked up in the upper reaches from a barnyard or a feedlot, can go unnoticed and, even if suspected or discovered, are hard to document and prove.

Pollution in both basic groups falls into eight categories, as a fouling agent of water, depending on source, makeup, and consequences:

1. oxygen-using sewage and organic wastes
2. infectious agents
3. toxic agents
4. various plant nutrients
5. thermal discharges
6. mineral discharges
7. sediment and silting
8. oil

Sewage and organic wastes that use oxygen originate mainly in domestic sewage, refuse from food- and dairy-processing plants, pulp and paper manufacturing, canneries, the making of textiles, and from large-scale poultry and animal farm operations. Because they rob water of its normal supply of oxygen, their effect on fish and other aquatic life is direct and too often lethal. They also foul streams in an especially offensive manner; some of the most objectionable instances of pollution I have seen, both as to appearance and smell, fell in this classification.

Infectious agents are disease-causing organisms, coming primarily from human waste, and from such industries as livestock slaughtering, commercial canning, dairy plants, and livestock production.

Toxic agents are poisonous chemical elements or compounds of many kinds. The list includes heavy metals, acids, alkalies, poisonous metals, ammonia, and cyanides. The chief sources are chemical industries and heavy manufacturing, particularly the steel and paint industries. A dramatic (and frightening) example of this type of pollution was provided by the mercury scare that sent shock waves through official health and conservation circles in many states and in Canada in 1970. In my home

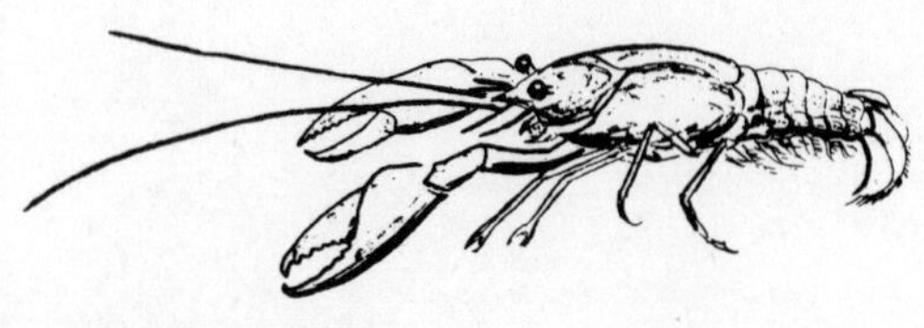

state of Michigan, the mercury problem exploded with the suddenness of a hand grenade. Mercury levels in fish caught in the St. Clair River and in Lake St. Clair, one of Michigan's most productive fishing areas, lying almost within the shadow of Detroit's skyscrapers, were suddenly found to be so dangerously high that the state's governor halted all commercial and sport fishing in the waters of the lake by executive decree as an emergency measure. The ban was later relaxed to allow anglers to catch fish, but they were forbidden to eat their catch.

Through the rest of that year, concern over mercury pollution mounted and spread across the country. Even rivers in roadless wilderness areas of Canada were found to be contaminated as a result of mercury wastes from paper mills on their upstream reaches. The alarm has subsided now but the threat remains; no one yet knows its true dimensions.

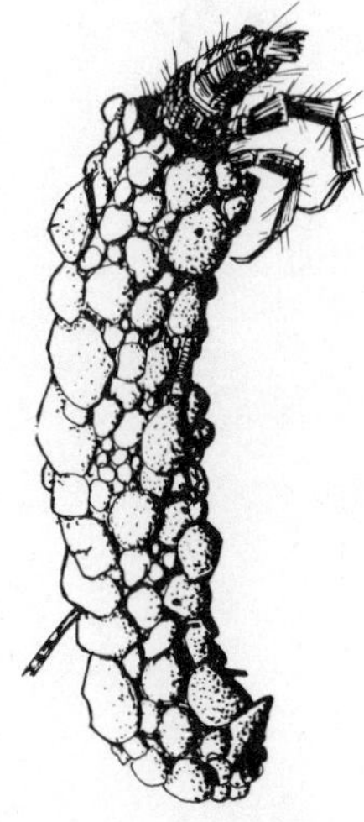

This same pollution category, toxic agents, also includes the very widespread and critical contamination of water by insecticides, pesticides, and herbicides.

Plant nutrients are organic or mineral substances in solution, used as food by aquatic plants. Detergents that contain nitrogen or phosphorous are the worst offenders in changing aquatic plant ecology. In many places the water of both streams and lakes is today so enriched by these substances that the growth of algae and other aquatic vegetation has been enormously speeded. As a result, the lake or stream is fated to die long before its time. Scientists call the process eutrophication. Phosphates from human homes are its most common cause, and disaster is its inevitable result.

As to thermal discharges, electric power plants, nuclear plants, steel mills, and petroleum refineries are the major offenders in this category. All use great quantities of water for cooling purposes, and it is close to standard practice to return this heated water to the river of its source. The resulting increase in temperature has serious detrimental effects on the aquatic food chain. It can even create a thermal "plume" in which fish cannot live and through which they cannot even migrate.

Mineral discharges are usually caused by the mining of coal, gold, sand, gravel, and other minerals, and their foremost threat results from strip or surface mining.

Sediment and silting come from many sources. Some sedimentation results from unavoidable natural causes, when storms and floodwaters wash soil from the land and deposit it in streams and rivers. Much more is preventable, however. Road building, logging, urban development and other human activities are the offenders, and lack of proper planning must take the lion's share of the blame.

Blame for some of the country's most disastrous pollution has to be laid at the door of oil companies. The sources are many and varied. Oil spills at sea have provided the most dramatic and disastrous examples in recent years, but the fouling of inland waters with oil is hardly less serious, although far less well known. Storm sewers carry some of this oil into streams, but direct dumping by industry accounts for far more. One of the most shocking examples occurred a few years ago when the Cuyahoga River, which flows into Lake Erie at Cleveland, became so fouled by industrial oil pollution that it actually caught fire and burned.

After water pollution, the stream killers that stand highest on the dishonor roll are channelization, dams, and the acid that leaches from the aftermath of unreclaimed strip mining. The latter is a form of pollution, of course, but its consequences are so deadly that it deserves to be discussed by itself.

I would put the three in that order.

Dams have wiped out more miles of stream and river than any man can enumerate, and fear of the dams of the future hangs like an ominous cloud over the quality of flowing water and stream fishing in the years to come, unless the dam builders are brought to heel. And acid pollution from surface mining has wrought damage that is hard to believe. But channelization is worse.

To date, this brainchild of the U.S. Army Corps of Engineers, recently adopted by the Soil Conservation Service, has gutterized and ruined more than eight thousand miles of stream in some forty states. There are plans for the future that will involve every state in the Union, including even Alaska. The National Association of Soil and Water Conservation Districts contends that nine thousand watersheds in this country are "in need of treatment and development."

If and when that huge program is carried out, there will be hardly a small stream left in the nation in its natural condition, and the further result will inevitably be additional channeliza-

Man decides to "improve" on nature's "inefficiency" by dredging and channelization. This is the result.

tion or more dams on the rivers into which those gutted creeks pour their floodwaters.

This nationwide process of gravedigging consists of dredging, clearing, and straightening small streams, creeks, and even major rivers. When the job is finished, miles of winding natural watercourse have been converted into a network of sterile and unsightly ditches. Natural cover along the banks has been scalped away and erosion accelerated. Swamps and marshes have been drained, silt-laden rainfall and snow-melt water sent hurrying downstream out of wetland areas, to create an excuse for one more dam on a major river.

Forgotten is the injunction that long governed land and water management: *Keep the raindrop where it falls.*

The consequences to fish and fishing are easily imagined. "Miles of our prime trout streams have been reduced to straight, wide, shallow, featureless ditches, with high dikes substituted for banks," says Ralph Abele, executive director of the Pennsylvania Fish Commission. "Stone, shale and gravel taken from the stream beds have been piled six to eight feet above the original banks. The only way a fish could ever travel these streams would be lying on its side!"

The example of the consequences of channelization I like best is Prairie Creek, a small stream that once meandered through swamp and wooded bottoms in Daviess County in southwestern Indiana. In the summer of 1970 a fishery biologist working for the Indiana Division of Fish and Wildlife put a shocking-boat on a pool there, intending to undertake a fish count. A channelization program in the 90,000-acre watershed was all but completed, and forty-nine miles of the creek had been reduced to shallow drainage ditches without bank cover. Conservationists were calling it the worst dredging project ever carried out in Indiana so far as fish and wildlife were concerned. Officials of the Soil Conservation Service and the Indiana Department of Natural Resources were touring the area that day, along with local conservancy-district people. The fish-shocking demonstration, in the only pool left in the forty-nine-mile stretch of stream, at a railroad bridge, was in part for their benefit.

The biologist worked the entire pool with his equipment. He turned up thirteen gizzard shad, seven high-fin carpsuckers,

Channelization is the essential ecological tragedy: this stream had established its own meandering course over a period of thousands of years, and once moved at a pace that supported life systems in harmony with the rest of the environment. *North Carolina Resources Commission*

three longear sunfish, two carp, two spotfin shiners, two silver-jaw minnows, one longnose gar, one bluegill, and one small-mouth bass. The people who lined the bank called it a tremendous haul.

The project that did the most to bring the ruinous potential of channelization into the open, and rouse fish and game officials across the country to angry protest, was the Alcovy in the counties of Gwinnett, Walton, and Newton in central Georgia. Announced in 1968, it was a gigantic scheme. It called for the dredging of eighty miles of stream, including a large part of the Alcovy River; it was to take seven years to complete and cost $9,000,000. Of that huge sum, the Soil Conservation Service would provide $7,000,000, leaving local backers a relatively small tab to pick up.

In a burst of rose-tinted oratory, an SCS spokesman made the prediction that the project would "forever remove the dire threat of floods, increase the prosperity of the whole region, and enrich the wildlife community."

Georgia conservationists called it a total tragedy.

With the backing of the Game and Fish Commission, federal conservation agencies and citizen groups, George Bagby, director of the commission, blew the lid off. The project would obliterate one of Georgia's most picturesque fishing and boating streams, the unpolluted Alcovy, he charged, destroy thousands of acres of prime deer habitat and waterfowl wintering grounds, and eliminate Jackson Lake as a place for recreation. Jackson was the most popular fishing hole near Atlanta and Macon, and the dredging of the river and its tributaries would threaten the lake with a year-round flow of mud and silt.

An investigation by United States Representative Ben Blackburn of Georgia revealed a shocking situation, too. Of 176 landowners along the Alcovy whose swamplands would be converted into cropland by the eighty-mile-long dredging project, 103 were already receiving "crop support" payments from the federal government for leaving farmland idle. Payments to individuals were running as high as $5,000 a year.

The controversy was bitter, a political battle fought without quarter. The project was blocked in part, but the agencies that dig the graves of living streams are powerful and tenacious, and the final outcome is still in doubt.

It is hardly a wonder that United States Senator Ernest

Swamp bottomlands that were to be drained by channelization along the Alcovy, in Georgia. *Luther Partin*

Hollings of South Carolina, battling an SCS proposal to ditch and drain the Horse Range Swamp in that state, at the end of 1972 called channelization "an environmental disaster and a proven mess."

No threat to the remaining free-flowing streams of America can match it at the present time.

The Army Engineers and SCS are too powerful, too politically entrenched, to heed the outraged outcries of private citizens. The remedy lies in one step and only one. Public Law 566, under which SCS operates, *must* be amended to give state fish and wildlife agencies and their federal counterparts the power of veto over any channelization project that would replace natural streams with shallow ditches, destroy fish and game habitat, do more harm than good.

In the meantime, however, state conservation agencies are not entirely powerless. In Louisiana, where hundreds of small streams and many larger ones have been converted into muddy ditches, the state legislature, acting at the urging of the Wildlife and Fisheries Commission, enacted a law that established a stream-preservation system on part or all of twenty-nine major streams, bayous, and rivers. The law insured that these waters will remain in their natural scenic condition. The headlong rush to destroy what is left has been checked in that state. And at the end of 1972 the Michigan legislature passed a tough Inland Waters Bill that vests complete control of all streams and lakes in the state with the Department of Natural Resources and prohibits any dredging or filling until the department has issued a permit.

Little by little the thrust of the gutter diggers is being checkmated by determined counterattacks.

When it comes to dams, small ones have done dreadful damage on countless small rivers, but it is the high dams on larger river systems that provide the most spectacular consequences of excessive dam building.

In the arid West in particular, the urge to dam streams seems close to uncontrollable. To many a westerner, water running freely down its natural channel is water going to waste. "Water development" is a magic label, to a degree that easterners find difficult to comprehend. And to ranchers, farmers, real estate developers, power interests, water officials and most politicians,

The improved channel of what was formerly a scenic, meandering bayou in Louisiana, whose shortcoming was what is known in engineering circles as an "ill-defined channel." That is, its waters spread out over the nearby permanent swamp, ridding itself of sediments and agricultural chemicals before it entered the food chain of the sport-fishing complex downstream. *Richard W. Bryan, Jr., Louisiana Wildlife Federation*

development means only one thing—dams. Preferably high dams, impounding big reservoirs.

The river whose death at the hands of the dam builders I remember most vividly—and with the deepest regret—was the Gunnison, on the west slope of the Rockies in Colorado. More than one authority called it Colorado's No. 1 trout stream, and nobody who had fished it ever forgot it. Some said it was among the greatest high-altitude trout streams in the world.

In June of 1959, when the hatch of the small variety of Dobson fly (parent of the hellgrammite), known there as the willow fly, was at its peak, I visited the Gunnison with Don Benson, a Colorado conservation officer. It was my first and last time to fish it. The river's death warrant had then been signed, by the dam builders of the United States Bureau of Reclamation.

I had never seen its equal as a trout stream where a fisherman could park his car on a paved federal highway, US 50, and almost step into the water. Plainly it was among the great rainbow streams of the country, as much a classic river in its own right as the famed Beaverkill and Willowemoc of the East.

"There'll never be another river that can take its place," Don Benson told me.

Today the pools and rapids and riffles, the wild, tumbling water of Black Canyon and the dirt road that ran beside it, all are drowned beneath reservoirs behind high dams at Blue Mesa and Morrow Point.

The fishermen-conservationists of Colorado lost their finest trout river by default. No major sportsmen's organization in the state spoke out against the dams; fishermen as individuals had little to say either way. "We were all opposed to the dams at first," the operator of a fishing camp told me, "but then we gave in. The idea seemed to spread that we couldn't lick 'em so we better join 'em and get what compensation we could for the damage that was going to be done."

How do you compensate for the loss of such a trout stream as the Gunnison?

The United States Fish and Wildlife Service and the Colorado Game and Fish Department (its director then was Tom Kimball, now executive director of the National Wildlife Federation) did all they could. They received virtually no public support.

The governor of Colorado branded a Fish and Wildlife Service report opposing the dams "ill-conceived, incorrect, misleading and incredible." The Game and Fish Commission, ignoring the objections of Kimball and his staff, gave its blessing to the dams. And on the public side of the fence, stockmen clamored for them, "regardless of the effect on fish and wildlife," as one horse-raiser's association said candidly.

It was small wonder the project went through.

In an article I wrote at the time for *Outdoor Life,* I gave this advice to fishermen everywhere: "Wherever you live, if a topnotch trout stream in your part of the country is threatened, remember the Gunnison. Speak up, shout from the housetops, fight, and do it in a hurry. If you wait for someone else to take the lead you may lose the river."

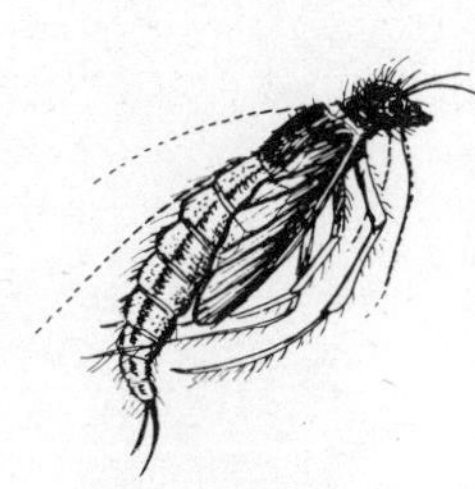

That is still my advice, more urgent today than ever; few streams like the Gunnison are left.

The results that can follow protests by groups of citizens acting in unison were dramatically illustrated in Michigan recently. The controversy involved the Grand, Michigan's longest river, which winds across the state for 260 miles, flows through many communities, and with its tributaries drains almost six thousand square miles. Nearly a million people live in the basin. Both the river and its major tributaries are fine fishing streams, with populations of smallmouth bass, northern pike, catfish and panfish, and substantial runs of salmon.

An organization called the Grand River Basin Co-ordinating Committee unveiled a plan for the construction of seventeen dams along the Grand and its chief tributaries to create a series of big impoundments. The Army Engineers would build the dams, in the name of flood control.

A brush fire of public indignation spread swiftly the length of the river basin, fanned by sportsmen, conservationists, and families who were to be displaced. The callous attitude of the proponents of the dams was revealed clearly when one of their spokesmen, a professor of urban planning, told a public hearing, "Southern Michigan is going to be an industrialized urban area. The natural environment has to play a secondary role."

It didn't work out that way. The governor of the state, William Milliken, entered the dispute on the side of the conservationists, urging that the natural course and flow of the river be maintained, and before the year was out the colonel who

Remember the Gunnison? Fishermen lost their finest trout stream by default. The United States Fish and Wildlife Service and the Colorado Game and Fish Department received virtually no public support in their opposition to the proposed damming of one of the finest trout streams in the country. These wild, tumbling stretches of the Gunnison now lie drowned in reservoirs.

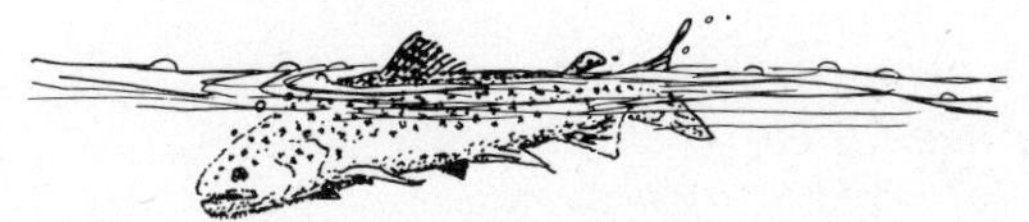

headed the Detroit office of the Army Engineers, and was also chairman of the coordinating committee, announced that the dam plan had been scrapped. "The reaction from the public was unfavorable," he explained. "The people didn't want those flood reservoirs."

The lesson of that victory is one to remember.

The classic example of the destruction wrought by big dams is the Columbia River and its major tributaries. That huge system, draining a sizable part or all of five states in the Pacific Northwest, plus a share of British Columbia, with a flow second only to that of the Mississippi among the rivers of America, was once among the most productive rivers in the nation for salmon and steelhead trout. Now it has been brought to its deathbed by dams built for flood control, power generating, and barge navigation.

There are today more than ninety dams, each impounding more than five thousand acre-feet of water, in the Columbia basin south of the Canadian border, plus many smaller ones for which no accurate count exists. And more are being built.

The major dams are massive and high. If the one proposed for the High Mountain Sheep site in the Hells Canyon of the Snake is finally built, it will soar an unimaginable 670 feet from top to bottom.

Those dams and their side effects killed *75 percent* of the smolt run (young trout and salmon making their first migration to the sea) in 1970 and *90 percent* in 1971. A Washington fishery biologist warned that three more years of such losses would mean the end of these fish in the upper river. "They will vanish into oblivion," said the Washington Game Department's fish scientists.

In addition to the formidable barriers posed by the dams and their turbines and impoundments, an even graver threat hangs over the river's fish populations: the little understood poisoning of water known as nitrogen supersaturation. Few fishermen have heard of it, but it is a deadly killer.

As water plunges over the spillway of a high dam, air is

Bonneville Dam, on the Columbia River just upstream from Portland, is just the first barrier that spawning runs of salmon and steelhead encounter when they enter the river.

trapped and carried down with it, deep into the stilling basin below. There the pressure of the falling water compresses the air, causing abnormal quantities of nitrogen to be dissolved. Fish exposed to these excessive levels suffer about the same consequences as human divers who rise to the surface too rapidly and suffer the crippling or killing affliction known as the bends.

For fish, the results are horrible. Bubbles of free nitrogen appear under the skin and in the fins, tail, and roof of the mouth. The eyes protrude or hemorrhage, and in extreme cases are actually blown out of the head.

Nitrogen supersaturation in the Columbia and Snake has reached a level where it threatens the survival of all fish migrating through or living in the water. The system has been converted into an ecological minefield.

The Columbia is an extreme example; there are far too many lesser ones to count. I do not know a state wildlife agency in the nation that does not cry, off the record but with bitter emphasis, "Damn the dammers!"

Next on the list of stream killers stands strip mining.

No one knows exactly the extent of the damage caused by this eyesore industry, but it has left its blighting scars on at least five thousand square miles of American land and seriously harmed not less than thirteen thousand miles of stream, one hundred three thousand acres of natural lake, and forty-one thousand acres of impoundment.

To understand this type of water poisoning, you have to know that above or below many coal seams lies a layer of rock or earth impregnated with iron-sulfur compounds that are a source of sulfuric acid. When that earth is exposed to air and rain by the process of surface mining, the deadly acid seeps out.

The Blackwater Falls State Park in West Virginia is a prime example. One of the most idyllically beautiful scenic spots in the eastern half of the country, until about 1960 that stretch of the Blackwater River was as famous a trout stream as any in the state. Then the fishing came to an abrupt end.

A few miles upstream from the falls and the wild gorge below them, two acid-laden creeks ran into the Blackwater from what had become strip-mining country. By the 1960s an official guide to the state parks of West Virginia listed Blackwater as the only park out of a total of twenty that had water, but no fishing.

This steelhead smolt shows symptoms of nitrogen poisoning, one of the lesser-known yet deadliest effects damming has on fish. The bubbles, most evident on the head and tail, are filled with nitrogen from the supersaturated waters below the dams. *Washington State Department of Game*

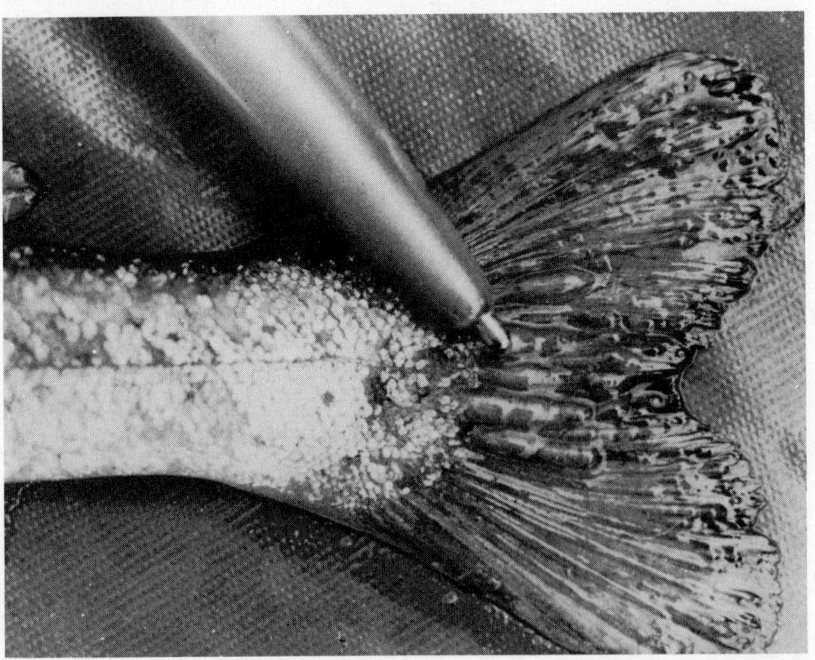

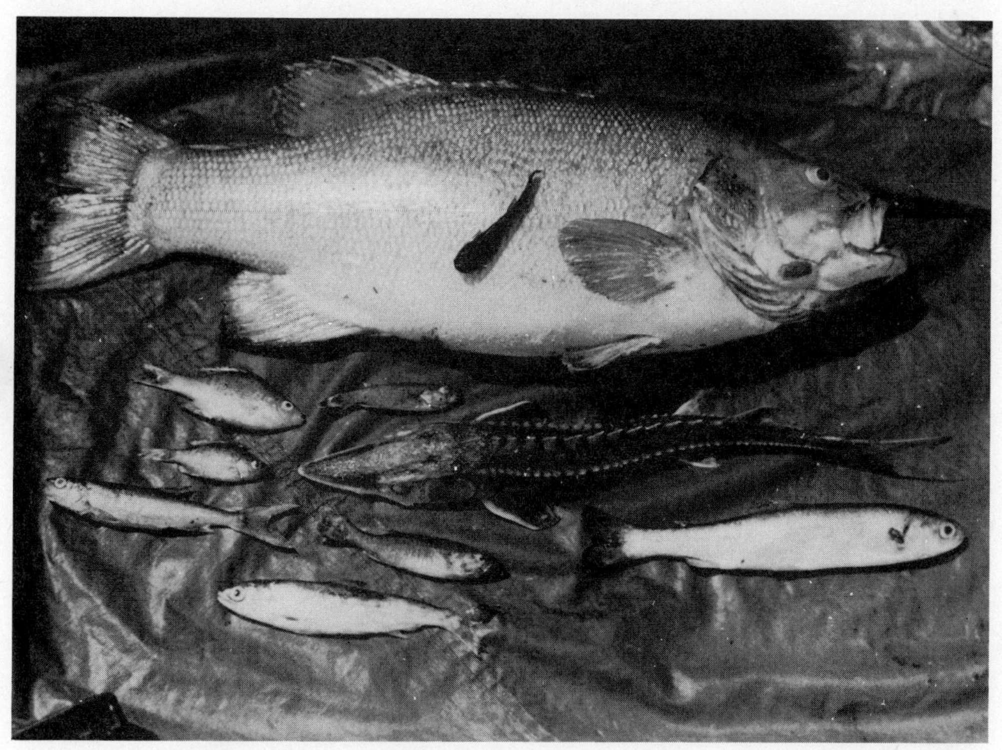

Salmon and steelheads are not the only victims of nitrogen poisoning. All these fish died the same way. *Washington State Department of Game*

West Virginia's Blackwater River and Falls in winter has a pristine, ghostly beauty that is all too revealing. Acid seepage from strip mines upstream have killed this river as a biological system: not one minnow, insect, or fish lives in this water.

The West Virginia Department of Natural Resources built the largest limestone-drum installation in the nation, hoping to neutralize the acid, but a flood washed out the dikes in 1972. When the system is fully operational, reclamation officials hope for "some improvement" in water quality—but, for now, that stretch of the Blackwater remains devoid of trout.

There is a small stream in the strip-mining country of southeastern Ohio that I saw only once but shall never forget. It wound pleasantly through wooded bottoms, with farmland and timbered hills on either side, and almost certainly it had been a smallmouth river at one time. But when I saw it in the sixties it was dead. Not a minnow swam in the shallows, not a dragonfly hovered over the water. Acid wastes had done their lethal work thoroughly. Neither livestock nor wildlife could drink from that stream.

Another major source of acid pollution is the slag heaps, mountainous heaps of refuse left when coal is cleaned. Consisting of impure coal, rock, pyrites, and other waste material, they cover acres of land like dirty gray scabs, and sulfuric acid can leach from them for as long as a century. More than three million tons of this material drains each year into the streams of Appalachia, and there is not a state in the coal-mining regions of the eastern mountains that has not lost streams to such poisoning. In Pennsylvania, for example, acid mine drainage is far and away the most serious pollution problem, but fish kills from this cause now occur very rarely—for the polluted streams are dead, and for years there have been no fish left in them to be poisoned.

Streams suffer less from pesticides than lakes, since running water has the ability to cleanse itself by washing the residues downstream. Also, the pesticide threat is somewhat less alarming today than it was a few years ago, now that the use of DDT has been phased out. But it still exists. In many states streams are damaged by the use of various pesticides and herbicides highly toxic to fish. Often the small-time polluter, the individual farmer (whose barnyard drains into a small stream or whose highly fertilized fields send their excess runoff into a creek) or the roadside spraying crew, is the offender. The wastes from one big farm feedlot equal the sewage output of a small

The terrible blight of dumping affects many Missouri rivers. *Don Wooldridge, Missouri Department of Conservation*

city, says the National Wildlife Federation, figuring on the basis of one cow to sixteen humans.

Here the individual fisherman can be of help, by reporting promptly to the proper authorities any case of pollution or fish kill he discovers.

The last two or three years have seen substantial strides made in the war against pesticide poisoning. In 1970, Wisconsin enacted a law banning the sale or use of DDT or any of its compounds, except under emergency conditions with prior approval from the State Pesticide Review Board. And that same year, Walter Hickel, then Secretary of the Interior, banned the use of sixteen of the poisons on federal land under the jurisdiction of his department, and sharply restricted the use of thirty-two others. Banned outright were DDT, Aldrin, 2–4–5T, Dieldrin, endrin, and mercury compounds, among others.

Yet for all these gains, at the end of that year Gene Gazley, now director of the Michigan Department of Natural Resources, sounded a sharp warning that the pesticide threat is far from ended and ripped the United States Department of Agriculture for its overcautious, foot-dragging approach to the control of so-called agricultural chemicals. "They are economic poisons, no longer confined chiefly to farms," Gazley warned. "It is not reasonable to leave their control in the hands of a department which has been deaf and blind to the rest of the environment." Pesticide manufacturers employ Madison Avenue sales tactics of half-truths, he added angrily, and the "we gotta spray" philosophy is mouthed to farmers by pesticide salesmen.

Thermal pollution in one respect bears a curious resemblance to pesticides: neither can be seen in the water, and consequently many fishermen-conservationists know little about them. Yet both can be deadly to fish life and the quality of a stream.

Hot water that, unless preventive steps are taken, is to be poured out from the huge thermal-nuclear generating plants of the future will spell the death of many rivers, major lakes, and estuarine and coastal areas. In 1968 the Atomic Energy Commission put out a map that showed a total of more than one hundred nuclear power plants scattered across the country. Fifteen were in operation, the rest under construction or planned for the immediate future. The map was updated late in 1972, and by that time the number of installations had grown to 146.

The Sport Fishing Institute predicted a few years ago that by 1980 the nation's generating plants, many of them nuclear, would require two hundred billion gallons of water a day, most of it for cooling. That is *one-sixth* of the nation's total average runoff. Even more ominous is the fact that, in heavily industrialized states, the *total* flow of the rivers involved may have to be passed through the cooling systems of power stations.

There is disagreement as to the consequences, even among

the experts, but many responsible authorities are hoisting storm warnings. For the most part, spokesmen for the power interests are disposed to minimize the threat, or shrug it off completely. Some of them say candidly and scornfully that power should be put ahead of fish, but scientists who study the environment, and public officials who deal with water quality, warn that if we tolerate thermal pollution on such a huge scale, we must expect the worst.

There are ways to avert this ruin. The water used in cooling can be cooled instead by passing it over huge ventilated towers. But these are costly, and to date the power companies have shown little voluntary interest in installing them, nor has any agency of federal government adopted a policy of forcing such installations. As a result of recent court decisions, however, there seems a fair chance that the Atomic Energy Commission will take a much tougher line regarding cooling towers in the future. But even if that happens, there is little likelihood that the towers will be required in every installation. Each case will be "decided on its individual merits" was the guarded prediction made to me early in 1973 by a spokesman for the commission.

Erosion is another prime stream killer. The most cataclysmic examples of this devastation usually follow channelization. Raw banks slide and wash, pouring their burden of sand, mud, and rock into the already sterile ditch. Lateral drains built by adjoining landowners add their load of soil, as do the cultivated fields that replace swamp and timbered bottomlands.

In the West, improper logging practices, especially clear-cutting of steep slopes and the building of logging roads close to streams, are often the major offenders. Along the South Fork of the Salmon River in Idaho, for instance, disastrous mismanagement of land in connection with logging, including careless public and private road building, has caused such ruinous silting as to kill that section of the river as a salmon stream.

Montana provides what is perhaps the nation's best example of a state's confronting a major erosion problem and discovering ways to deal with it. In the late 1950s the Montana Department of Fish and Game realized that the construction of the Interstate Highway system across the state posed the most serious immediate threat to trout streams. The fisheries division documented the extent of the damage on thirteen streams, and the

Severe logging practices, such as that on this hillside in Montana, cause untold damage to rivers.

department took its unhappy findings to the public.

The Montana Junior Chamber of Commerce and the Western Montana Fish and Game Association carried the battle to the legislature and the result was the Montana Stream Preservation Act, which requires that any agency or subdivision of state government must heed the advice of the Fish and Game Department on any construction project that involves a stream, with the sole exception of irrigation head works. There is still a loophole, in that the law does not apply to private individuals. It's a gap that sportsmen's groups are striving earnestly to plug, but the worst of the potential harm has been averted.

Montana still provides sorry examples of the effects of erosion from private road building, however. Logging roads are the main offenders. Rock Creek was one of the state's blue-ribbon trout streams a few years ago. It is a small stream, rising in spectacular alpine country west of the Continental Divide near

Missoula and flowing fifty-five miles to empty into the Clark Fork River. Many fishermen have regarded it as the finest trout water in the state on the western slope of the Rockies. A gravel road parallels it for much of its length, making the fishing easy to get to, and until recent times the water was pure enough to drink. But much of the logging that is being done in the two national forests that lie in the creek's drainage basin is clear-cutting. Timber and undergrowth are stripped away and the steep mountain slopes laid bare to erosion. Logging roads, often only 100 to 200 feet apart, wind across the denuded slopes in ugly parallel scars. In addition, many miles of road outside the cutover area inevitably go with such logging, and the road system accounts for as much as 90 percent of the sediment that is threatening Rock Creek.

Of all threats to streams, probably the most difficult to control is human use. This broad category of evils embraces irrigation, real estate development, urbanization, the sale of streamside lands for cottage or homesites, the use of off-the-road vehicles along roadless stretches of creeks and rivers.

In every state in the eastern half of the country, and to a lesser extent even in the mountain states of the West, the amount of publicly owned stream frontage is totally inadequate. As more and more of such frontage is taken up for real estate development or purchased for private dwellings, the stream suffers and access for fishermen is made more difficult.

"Today everybody seems to want a place on a river," comments a Missouri authority, "and most everybody is getting what they want." Michigan's recently enacted Inland Waters Law is designed specifically to prevent the spread of wildland slums that have already blighted many streams.

Fishermen, canoeists, and cottage owners compete on Michigan's famed Au Sable River. The newest—and perhaps most difficult problem facing the stream conservationist: overuse. Like any functioning system, a stream has a "load limit," after which its ability to regenerate itself becomes overbalanced by the demands made upon it as a natural system.

Before the logging era, Michigan's legendary Au Sable was a grayling river so famous that it lured anglers from the cities of the East Coast. In the last fifty years it has stood high among Michigan's great trout streams. Today it is so crowded with canoe traffic, and use conflicts among canoeists, fishermen, campers, and riverside property owners have built up to such a point, that many trout fishermen stay away and the river itself is under grave threat. Vandalism, trespassing, littering, and rowdy behavior reached an unbearable level, and in the fall of 1971 the Natural Resources Commission adopted a permit system for boats and canoes, together with a strict code of rules to regulate river use.

The code was immediately blocked by court action on the part of canoeing groups, however, and as this is written the issue remains unresolved and the Au Sable still stands as a classic example of a river that too many people want to use simultaneously for too many purposes, and that a rowdy minority seems determined to ruin. The DNR is now asking the legislature for clear-cut blanket authority to license and set quotas for marinas, boat and canoe liveries on all of Michigan's inland lakes and streams.

Some of the most splendid running water in the nation, priceless both for scenery and fishing, has been fouled, channelized, dammed, littered, ruined, while an indifferent people looked on.

The death penalty for flowing water continues to be imposed. There is more concern for the consequences today than ever before, but the forces that oppose it—both the new breed of conservationist worried about man's environment, and the fisherman who was crying out against stream desecration long before the word ecology found a place in everyday conversation—are not making headway fast enough.

A typical North Carolina river before channelization. *North Carolina Wildlife Resources Commission*

Enough said? *North Carolina Wildlife Resources Commission*

4

Streamside Surveillance

DON ECKER

Reverence for nature is compatible with willingness to accept responsibility for a creative stewardship of the earth.

RENE DUBOS

An industrial plant discharges toxic waste into a nearby stream and hundreds of trout, minnows, aquatic insects, and plants die. A town's antiquated and inadequate sewage system pours its semitreated or even raw sewage into a once-clear and clean creek—and the natural population of smallmouth bass that once thrived there are gone, with only suckers and bullheads able to survive. A chemical plant draws water from a neighboring river, passes it through its manufacturing system and returns it to the river ten degrees hotter than its original state.

"So what? Can't stop progress."

"Have to produce what people want and need."

"Got to put the sewage somewhere."

"So what if a few fish die? What's the river there for, anyhow, fish or people?"

These are some of the typical situations and excuses that confront the streamside conservationist. And if the conservationist is a fisherman as well, particularly a trout fisherman, he is more often than not accused of having the selfish motive of wanting to preserve his own favorite sport to the detriment of the population as a whole. It cannot be denied that the angler-conservationist has some selfishness in his motivation to fight to save his favorite fishing stream or river. All others should be eternally grateful that this is true. Were it not for the intensely dedicated activities of certain anglers and angling organizations, many great streams and rivers would now be totally defiled.

Many nonangling conservationists play an equally vital role in guarding our precious and irreplaceable watercourses, and there are many examples of general citizen action that has saved rivers from destruction. But the trout fisher is a born stream watcher. He patrols the banks of his favorite streams for almost seven months of the year, and many anglers I know make several trips to their best-liked rivers during the winter months, "just to see how things are going." Throughout the spring freshets, low and clear summer periods, and the varied moods of fall, the confirmed angler watches, wades in, and studies his stream. He knows the insect and plant life along the banks and in the stream itself; he is keenly aware of the fish population in the stream and quick to note any changes in their moods or habits. He studies the quality of the water. It is therefore logical that the angler assume the role of citizen protector of our watercourses.

This chapter will provide one approach to the detection of pollution that can be utilized by nonprofessional citizens, and will show how this technique is being employed successfully by particular groups of angling conservationists in cooperation with the federal and state governments. It is the natural and ideal adjunct to the vigilance of all serious and concerned anglers while fishing on a river.

Human, animal, industrial, agricultural, and physical pollution may be occurring on a particular watercourse at any time. The streamside conservationist may be able to detect that some of these pollution problems exist by simple, close, regular observation of the water—and his early detection may prove decisive in stopping the pollution before it becomes uncontrollable. Unnatural changes in temperature, color, odor, degree of turbidity, aquatic plant growth can serve as indicators of possible pollution. Other visual observations, such as the presence of floating solids or suspended matter, dead or dying fish or other aquatic fauna, also serve as important warning signs of pollution. However, in order to assemble the information gained through observation and relate it to a plan of action, a

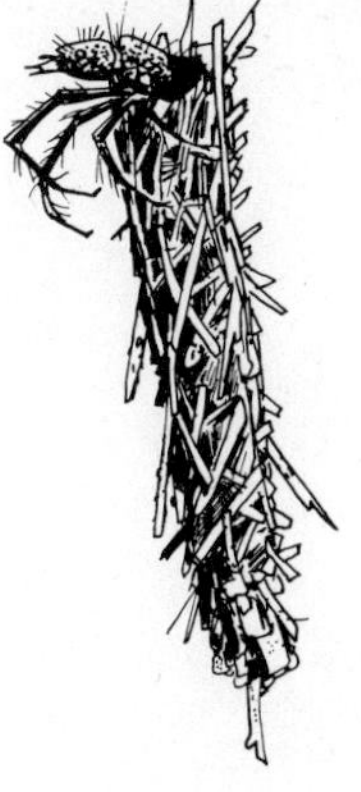

degree of organization and standardization must be achieved.

A most effective "early warning" system has been developed by the Theodore Gordon Flyfishers of New York City, in cooperation with the New York State Department of Environmental Conservation (DECON). Before describing the program itself, it is worth showing briefly how the Water Watchers got under way, since this may be helpful to other clubs, groups, and individuals across the country.

In 1970, Phil Chase, a TGF member and director, suggested that the group institute the program now called "Water Watchers." Phil had been conducting coliform bacteria tests on several streams in the Catskill Mountains—principally the once-famous Neversink and its tributaries, which had been one of the home waters of Theodore Gordon. Phil was most disturbed about the abnormally high bacteria counts he was experiencing, and he presented this data to a TGF gathering. It was suggested that such stream testing might be an important safeguard for *many* streams throughout the state, if an organized program could be instituted. Even more important, the question was raised whether DECON might be interested in working with TGF to develop such a plan.

TGF Conservation Chairman Gardner Grant went to Albany to discuss the matter with DECON officials. At first there was resistance and a good deal of scepticism over the idea of a volunteer citizens' group doing an effective and continuously reliable job of stream testing and monitoring. DECON personnel were already doing some sampling, and together with industry and community representatives were collecting data from two hundred sites in the state, plus a dozen automated monitoring device locations. DECON officials were worried that after the first rush of enthusiasm had worn off, the TGF volunteers would rapidly lose interest in the repetitive and somewhat inconvenient procedure of testing every month, particularly throughout the cold winter months. But despite these reservations, DECON gave the project a provisional green light.

A program was designed for testing water quality by citizens with little or no prior technical knowledge or training. The four parameters selected for testing and reporting were:

1. coliform bacteria

2. dissolved oxygen (DO); oxygen analysis in good trout streams usually indicates from 4.5 to 9.5 parts per million

3. acidity-alkalinity (pH)

4. temperature (water and air) plus *all relevant visual observations.* Highest limiting temperatures: brook trout, 75° F, brown trout, 81° F, rainbow trout, 83° F

DECON personnel worked with TGF in selecting the appropriate equipment and determining which streams would be monitored. Testing sites were selected by agreement between DECON's Chief of Water Surveillance, Ron Maylath, and the captains of the monitoring teams that had been formed by TGF. In all, five teams were set up, consisting of five or more members each. The plan was to obtain samples and data from six separate locations in each assigned area, on a once-a-month schedule for each team. Copies of the test reports would then be forwarded to DECON's local field office and an additional copy to the DECON Water Quality Surveillance office in Albany. Any reports of emergency situations could be acted upon quickly by the field office people, and in Albany the accumulated data would be placed in the state's computer memory banks. A water-quality profile could then be built up for each testing area, and long-term trends evaluated. This accumulation of water-quality data would help to establish a basis for checking on whether regulations and "cleanup" orders were being followed, and on the need for possible changes in existing regulations.

Other copies of the monthly reports would be sent to both TGF's Project Director and Conservation Chairman, so that the teams' work progress could be checked. In any such government-citizen joint undertaking, public relations play an important role. To further the public awareness of the program, TGF had special permanent aluminum signs made and posted at every test site. This would also serve to assist any new testing members to locate the precise assigned locations for sampling.

The Water Watchers program now needed the recommended testing equipment—and trained testers. The Hach Chemical Company, Inc., of Ames, Iowa, was selected as the source of the testing kits for the DO analysis and the pH determinations. (These tests must be performed at each testing site, immediately upon sampling.) The coliform bacteria testing equipment was obtained from the Millipore Corporation of Bedford, Massachusetts. (Coliform bacteria are present in the waste of humans and other mammals; thus, this test is a vital measurement of water quality as it provides an indication of contamination by sewage, one of the major pollution problems throughout the country.) By federal standards, water which has more than 1 coliform per 100 milliliters is unsafe for drinking. The Amawalk, which is one of the TGF test streams and generally considered a fine piece of trout water, therefore relatively clean, averages as much as 100 coli for each 100 milliliters.

After reaching agreement with DECON, a training program was created using a concisely written manual that could be understood easily by any team member whether or not he or she had any previous technical experience. This manual was augmented by a slide presentation that Phil Chase prepared under the auspices of the Federation of Fly Fishermen. After a series of training sessions, which included lectures by Maylath and other DECON officials and personnel, the teams went into the field to begin their sampling, testing, and reporting.

A shakedown period was necessary to work out the many "bugs" that necessarily occur in such a first-time program, but as soon as the teams got into the routine of their work, they accumulated much valuable data. The attitude of DECON, after the Water Watchers program got into full operation, can best be described by a statement from Ron Maylath. "We are behind them 100 percent," he said. "TGF has an active, constructive program and is not just paying lip service to the environmental effort."

HOW WATER WATCHING WORKS

The actual field-testing procedures are relatively simple; with a minimum amount of training, almost anyone with a high school background should be able to achieve some degree of proficiency within a short while. The equipment is so designed that reliable and accurate data can be secured by people with little or no special scientific training.

In setting up testing locations, it is important to select sites along any particular watercourse that can pinpoint possible problem areas—where pollution may originate. The testers should take samples at each site at least once a month; more often would be desirable, but experience has shown that if too great a work load is placed on the testing groups continuity may suffer. It is best to set reasonable goals—and stick to them.

The sample form, labeled FFF#1, has been developed so that a nonprofessional testing team can indicate with a reasonable degree of accuracy, and in a uniform manner, the results of their observations. It is a simplified form, developed specifically for use by volunteer personnel, and it has also been designed for easy extracting of the information to be used in data-processing equipment, so that computer profiles can be developed for each river under surveillance. Ideally, after a sufficiently long-term profile is established, any significant deviation from the normal ranges of the various water-quality factors will be immediately recognized by the computer as a danger signal; the government agencies having jurisdiction over the particular river are then alerted to potential problems, and can take remedial action should these problems be confirmed by more intensive testing by their own water-quality evaluation teams.

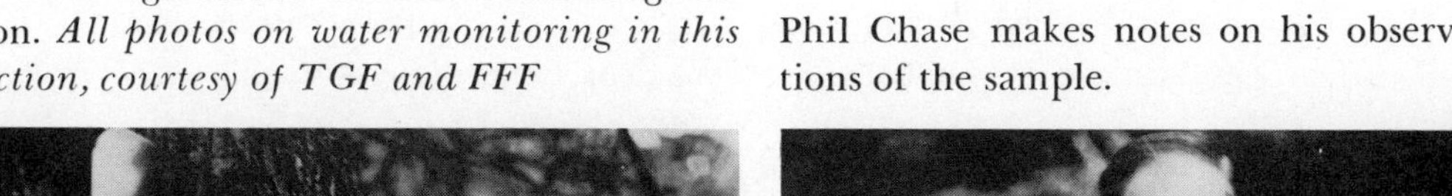

A TGF sign marks a water-monitoring station. *All photos on water monitoring in this section, courtesy of TGF and FFF*

Phil Chase makes notes on his observations of the sample.

To understand the sample form, let's run down a list of measurements and observations, with some general explanations concerning their relative importance.

Item 1. Weather. This gives a general indication of the weather conditions at the time the samples were taken—whether there was much rain or snow. Test results might be affected because of the dilution of the stream from the rain or melted snow.

Item 2. The percentage of cloud cover gives an indication of the amount of solar heating of the water. Higher rates of growth of bacteria and algae occur in warmer water, and the amount of oxygen that can be retained in the water decreases at higher temperatures.

Item 3. Air temperature provides an indication of the relative temperature of the air in relation to that of the stream.

Item 4. Flow. Measurements taken in moving water differ significantly from those taken in slower or still waters.

Item 11. Water temperature must be taken into account as an important measurement in the DO and coliform tests.

Item 12. pH. A stream that is chemically neutral has a pH value of 7. If a pH indication is less than 7, then the water is chemically acid. Higher than 7 would indicate an alkaline or basic stream. The more productive trout streams, such as the highly productive limestone streams of Pennsylvania and the abundantly rich and trout-filled spring creeks of Montana, have a generally basic or alkaline nature. Some of the eastern freestone streams tend to be more acidic; the typical brook-trout streams of the northeastern United States and the Pacific Northwest are examples of these types of waters.

Item 13. Dissolved Oxygen. As mentioned before, this is one of the basic tests of water quality. A natural stream of good quality usually tests out at higher than four parts per million of dissolved oxygen; a lower test result is an indication of serious water-quality problems.

Item 14. Determination of biochemical oxygen demand (BOD). Basically, this is the demand placed on the available oxygen by the organic matter in the stream, which requires oxygen for decomposition. The BOD determination is a measure of the weight of dissolved oxygen consumed in the abovementioned biological process in a particular sample of water. Since a scientific determination of BOD requires a number of

FFF WATER QUALITY SURVEILLANCE PROGRAM - RESULTS OF EXAMINATIONS

Lab. No.		Project No.
Collected by:		Date rec'd. at Lab.

51. Dr. Basin No. (1 2)	52. Station No. (3 4 5 6)	53. Date Collected (7 8 9 10 11 12 — Month Day Year)	54. Time Collected (13 14 15 16)

1. Weather (1, 2, 3) — 17
   1 = Clear
   2 = Overcast
   3 = Precipitation

2. Percent Cloud Cover — 18 19

3. Air Temperature (°C) — ± 21 22 23

4. Flow (cfs) — 24 25 26 27 28 29 30

11. Water Temperature (°C) — 53 54 55

12. pH (units) — 56 57 58

13. Dissolved Oxygen (mg/1) — L/G 60 61 62

14. B.O.D. (days) — 63 64
    No. Days of B.O.D. test:
    5-days, 7-days, 21-days, 28-days, etc.

15. B.O.D. (mg/1) — L/G 66 67 68 69 70

57. [78] 1 58. [79] 1 59. [80] 1

17. Coliform (test) — [17] 3

18. (no./100 ml) — L/G 19 20 21 22 23 24 25 26

57. [78] 1 58. [79] 1 59. [80] 2

30. Phosphates (mg/1) (TOTAL) — 42 43 44 45

57. [78] 1 58. [79] 1 59. [80] 3

STREAM APPEARANCE

130. Color (code & intensity) — 17 18 19 20 21 (Code Code Int.)

131. Odor (code & intensity) — 22 23 24 25 26 (Code Code Int.)

132. Turbidity (intensity) — 27

133. Suspended Matter (intensity) — 28

134. Floating Solids (code[1] & intensity) — 29 30 31 (Code Int.)

135. Oil (intensity) — 32

136. Algae (intensity) — 33

137. Weeds (intensity) — 34

138. Biological Growths (intensity) — 35

139. Gassing (intensity) — 36

140. Foaming (intensity) — 37

BOTTOM

141. Type bottom (code)[2] — 38

142. Sludge (intensity) — 39

143. Sludge Thickness (feet) — 40 41 42

12. Stage Height Reading (feet) — 64 65 66 67

127. File Transaction — [78] 1

128. File No./ 129. Card No. — [79] 6 / [80] 9

1. 1 = Sludge
   2 = Small Solids
   3 = Large Debris

2. 1 = Rocky
   2 = Sandy
   3 = Gravel
   4 = Mud
   5 = Sludge

REMARKS:

Form FFF-1 FFF — WATER WATCHERS

days to complete, the programs discussed in this chapter—which are being conducted by various volunteer groups—do not, at this time, include BOD determination. It is hoped that these measurements will be added at a later time—for this is an important determination for any water-quality evaluation. Streams with a high BOD measurement are considered polluted. Quality water, and particularly good fishing water, generally has a low BOD reading. This test directly pertains to waters subject to the addition of sewerage effluent.

Item 17. The coliform determination is discussed elsewhere in this chapter in greater detail, and of course is also related directly to sewerage problems.

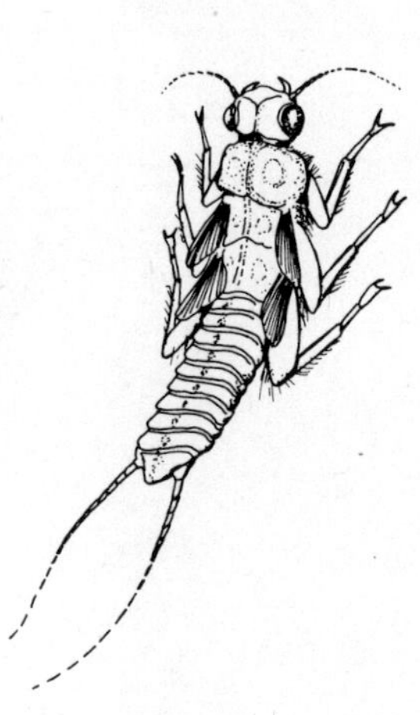

Item 30. Phosphates. The continued use of modern detergent substances by household consumers and industry has placed a heavy burden on the rivers of our country. Even relatively up-to-date municipal treatment plants often do not include a process for the removal of phosphates. When high concentrations of these substances are added to our rivers, one of the principal long-term negative results is overenrichment, with resultant increases in the growth of aquatic plants and algae. Over the years, this increased growth of organic matter—with its eventual decay as part of its life process—creates a situation known as eutrophication. This is a process whereby a sludge or sediment is created from the decay of organic matter in the stream; it fills up the bottoms of pools and the slower areas of the stream bed, eventually changing the nature of the stream bed itself.

During the decaying process, harmful gases are formed that are sometimes toxic to fish and other aquatic fauna and are most unpleasant to those people who use the stream for recreational purposes. As this condition occurs, the DO level drops and the BOD value climbs. At the present time, phosphate determination is also not included in the FFF Water Watchers program because of the technical nature of this determination. (Tests for mercury contamination are not included for the same reason.)

Items 130 through 140. These are gross observations made visually at the time the samples are taken. A code has been developed for the sake of good order and ease of reporting.

Dissolved Oxygen Analysis

The Hach Chemical Company's kit includes a water sampler with a carrying case and the test kit itself.

One sample obtained with the water collector can be used to do the DO, pH, and coliform tests. Prior to taking the sample, note the air and water temperatures, for the DO test is otherwise not valid. Notes should also be made of the relative height of the stream; the ideal place for this would be at a bridge site, where the relative water level can be marked on the bridge footing. One properly located gauging station is adequate for a stream. The sample is collected according to a prescribed procedure supplied with the kit.

The glass-stoppered bottle is filled with the water to be tested. The water should overflow the bottle for several minutes to remove any air bubbles. The contents of one prepared package of DO Powder #1 (manganous sulfate) and the contents of one prepared package of DO Powder #2 (alkaline iodide-azide) are added to the sample bottle. The bottle is then stoppered carefully, to avoid trapping any additional air. Next, shake the mixture thoroughly. A precipitate will appear, which should be brownish orange if there is oxygen present. The bottle is then set aside until the precipitate has settled halfway.

Now shake the bottle again and allow it to stand until the upper half is clear. The contents of one package of DO Powder #3 (dry acid) is then added, the bottle carefully stoppered again and shaken thoroughly. The precipitate should then dissolve and yellow appear, indicating the presence of oxygen. This is the prepared sample, ready for the determination of dissolved oxygen content.

Now take a measured sample from the bottle and pour it into the mixing bottle. Add a titrating solution (PAO), drop by drop, while swirling the sample to mix. Each drop should be counted until the sample becomes colorless. Each drop used equals one part per million of dissolved oxygen.

Should the results from this determination be on the low side (3ppm or less), a larger sample is taken to get a more sensitive reading. This can be done by pouring off the contents of the DO bottle until the level reaches a mark indicated on the sample bottle. The titrating solution can then be added in the same fashion as before, counting each drop until the liquid

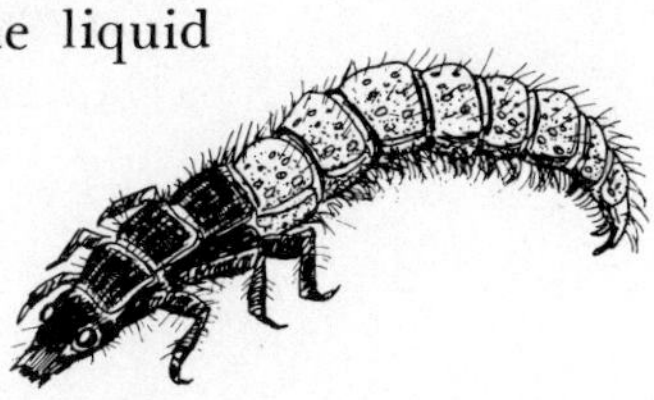

Taking a dissolved oxygen sample.

changes from yellow to colorless. Each drop added is then equal to 0.2 ppm of dissolved oxygen. The results from this determination are then noted on the FFF #1 form, or another appropriate form.

Acidity-Alkalinity Analysis (pH)

This simple test relies on a color comparison. Two color-viewing tubes are filled with the samples. Six drops of a pH indicator are added and mixed in one of the tubes. The two tubes are then inserted in a color comparator device that is then held up to a source of light, such as the sky or a lamp. By rotating a color disk, a color match can be achieved. The pH

Adding the titrating solution to the dissolved oxygen sample.

Taking a water sample to determine the stream's acidity.

level is then read directly from the scale on the comparator. If chlorine is present in the water in greater concentration than one part in a million, it can be removed with a dechlorinating solution. The results from this determination should then be recorded.

Coliform Bacteria Analysis

This analysis consists of three parts, only the first of which need be done at streamside; the second two parts can be done in a home workshop, basement, or garage, but should be accomplished as soon as possible after the sample has been taken.

There should be sufficient water collected in the water sampler to provide for the DO and pH determinations as well as the coliform testing.

A convenient and inexpensive container, which has proven most useful, is the common Styrofoam coffee cup that holds about ten ounces and has a tight-fitting plastic cover. The sample should be poured into the coffee cup and then covered and maintained at a temperature as close to the temperature of the stream as possible.

The test consists of filtering the sample through a device that permits the passage of water but will not permit the passage of any coliform bacteria that may be present. The Millipore Kit has been selected by TGF, FFF, and TU (Trout Unlimited) as a simple and efficient setup for this test. The sample is poured through a filter with a vacuum formed beneath it. The filter is then placed in a petri dish containing a nutrient medium.

All procedures must be accomplished in a sterile manner so as to prevent outside contamination and allow an accurate count of the coliform bacteria colonies present.

The prepared sample is then incubated at 98° F for about twenty-four hours. At the end of this time, the number of shiny greenish colonies can be counted; the count is recorded in terms of 100 milliliter samples. The results are then noted on an appropriate form.

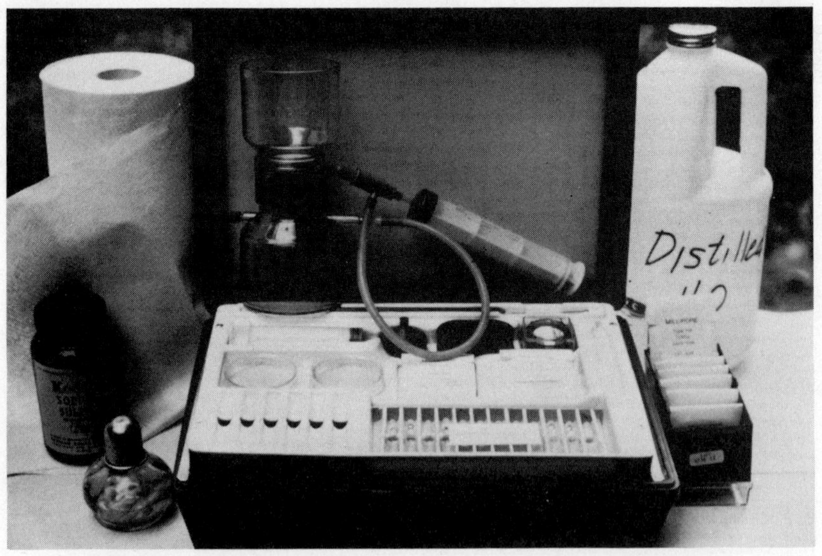

Millipore apparatus—used to determine the coliform count.

Taking a measured sample for the coliform count.

Coliform colonies after the incubation period.

Taking the stream's temperature.

This is a brief description of the procedures currently being used by a number of testing teams. Such stream monitoring can most effectively be conducted when it is part of a larger program carried out with the cooperation of a municipal, state, or federal agency, which can utilize the data derived and take appropriate action when problems are noted.

A particularly valuable side benefit of the Water Watchers program is that the information and experience gained through regular testing can be put to use in public debates and hearings on water-quality issues. The more informed those devoted to clean waters are—and armed with specific, detailed information, perhaps gathered themselves—the more effective they will be.

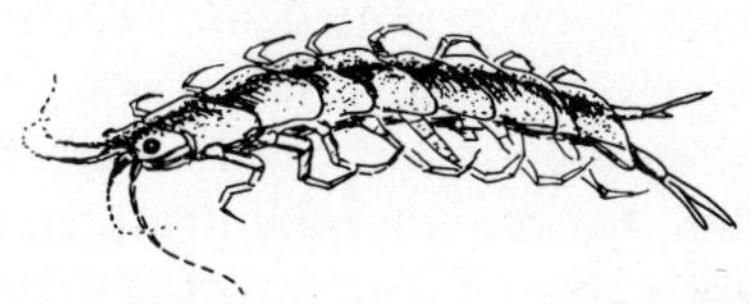

GOVERNMENT AND CITIZENS— A PARTNERSHIP

The great importance of the TGF program, beyond the generation of information on water quality, lies in the fact that it represents a *partnership* between government and a citizens' organization. It is estimated that the work performed by TGF monitoring teams represents a contribution of at least $18,000 a year in labor costs alone. In addition, TGF finances the purchase of all their own testing equipment, at a cost of as much as several thousand dollars a year, and also covers the cost of all consumable testing chemicals and materials. State groups have neither the manpower nor the funds to monitor more than a few major rivers, and thus such a program provides vital computer input otherwise not obtainable. The cheapest and most effective way to achieve pollution control is to *prevent* a river from becoming polluted in the first place; not only does the TGF plan help to track down sources of pollution, but it aims at spotting problems *almost as soon as they appear*—and thus seeks to prevent the slow, almost unnoticeable decline in the water quality of a river.

With the initial success of the Water Watchers, the next logical step was to expand the number of teams and test sites, and hopefully to extend the program into neighboring states. Manpower is the key to such expansion, and TGF did not have to look far to find kindred spirits in the fight to protect our trout waters. The logical new partner was Trout Unlimited. TU's particular interest is in the maintenance of the proper pure water conditions and habitat for the optimum biological level of a healthy trout population, and, wherever possible, a *natural* trout population.

Many cross contacts already existed within both TU and TGF, and soon various local TU chapters began to organize teams of Water Watchers to work within the established program. These new teams were supplied with the necessary equipment, material, training, and supervision.

An important development occurred when Gardner Grant contacted the Federal Environmental Protection Agency concerning their possible national support of the Water Watchers. This agency showed definite interest and suggested that a particular team might be set up under joint TGF, TU, and EPA auspices to serve as a pilot project team for future national expansion and funding of the program. Under Fritz Gerds, president of the North Jersey Chapter of Trout Unlimited, a monitoring team was formed and this group received its initial training and equipment from TGF, with additional training from EPA itself. They were then assigned to test the Passaic River, selected because it is so highly polluted that the gathering of data on its water quality would put the team to a strenuous test and determine its reliability and testing capabilities.

The final decision of the EPA concerning expanding and funding the program on a national scale is still pending; but there is every hope that their decision will be favorable: it would give this practical program of stream vigilance great impetus.

There is nothing exciting or glamorous about water watching. To be of significant value, a monitoring program relies on discipline, dedication, and hard long-term commitment. But anglers are a dedicated lot. They are steeped in the traditions of the past and respect for the future. Such men have proved that they can religiously gather data that may have no apparent immediate use, confident that they are making a contribution to the present and future enjoyment of the outdoors and the maintenance of clean waters. They also know that they will be gaining a greater understanding and appreciation of their favorite stream than they ever had before.

How can your organization begin its own Water Watching program?

Briefly, the steps you should follow are as follows—though you should start only if you are willing to persevere.

1. Select the watercourses you wish to monitor (perhaps with suggestions from government agencies in your region);

2. Choose testing sites and designate them on the appropriate USGS maps;

3. Organize teams, with a captain for each group (the suggested minimum is two, at least five is best);

4. Train and equip the teams as thoroughly as possible (see the TGF and FFF addresses at the end of this book, and write for specific contacts; they can now provide a complete audiovisual slide presentation and instruction manual with sufficient information to train teams and purchase equipment;

5. Appoint an overall project director who will supervise the work and coordinate the group with appropriate municipal, state, or federal agencies.

When discussing the pollution of rivers, an analogy is often made between trout and the caged canary miners used to take with them into mine shafts to detect the presence of deadly gases. Just as the highly sensitive canary dies from exposure to a harmful atmosphere, long before a miner may be overcome by the same fumes, trout also die or disappear in polluted streams well before such pollution ever has a direct effect on the nearby human population.

In the case of trout, however, the changes may be far more subtle and insidious. Trout may not actually die en masse, but the population may slowly diminish. This begins with the failure of the stream to replenish its stock of fish, as it becomes

unfit for supporting natural reproduction. The nymphs, larvae, and crustacea that normally inhabit a healthy stream also die off with the encroachment of pollution, drastically reducing the amount and quality of food available to support trout or other fish. Often other species, such as suckers and sunfish, will then begin to increase in numbers and compete with the trout for the available food. These species can tolerate a higher level of contamination than trout, though if the level of pollution continues to rise, these fish too will eventually diminish in number and disappear.

How should one report fish kills when they are first seen?

In cases of severe pollution, such as might occur if an industrial plant suddenly discharged a large and toxic quantity of poisonous waste into a stream, disastrous and dramatically swift fish kills may result—and they should be reported at once. Should you happen to be on a stream when a fish kill occurs, it should be reported immediately to the nearest field agent in your state department of conservation, Division of Fish and Game, or other agency having jurisdiction over water quality. To provide the most effective information, the following steps should be taken.

1. Collect evidence. A water sample is most important, taken in a clean container (not one that has been washed in the water affected by the fish kill);

2. Make a note (written if at all possible) of all physical conditions existing at the time: air and water temperature, strange odors or color in the water, insect life changes, silt or slime conditions, floating solid matter, scum or oil on the surface, and so forth;

3. Collect specimens of the affected fish. The best specimens are the freshest ones. A fish in distress or dying is the preferred type. If all fish are dead, the ones with the brightest red gills have died most recently. Keep the fish whole and cool;

4. Estimate the number and species of fish killed and record any other pertinent facts, such as "effluent coming from pipe; fish dying below it, not above";

5. Phone the local conservation officer or other appropriate field official;

6. Upon returning home, wrap the fish in aluminum foil or freezer wrap and freeze pending collection of evidence by the appropriate agency.

Pesticides accounted for this fish kill. Trout—like the miner's canary—die first; but panfish and coarse fish soon follow.

Pollution prevention, control, and abatement are vital necessities to everyone, angler and nonangler alike. Clean water must be the concern of everyone. The angler, however, is in the fortunate position of being constantly close to the watercourses of his area, and thus is able to play a unique and major role in the protection of his streams. By careful surveillance, he can help keep the water pure enough to support a thriving trout population; and while he enhances the enjoyment of his own sport, he also benefits the population as a whole.

Trout water is relatively pure water—but if the trout population cannot survive, like the miner's canary, what will be the eventual fate of the human population?

5

Stream Improvement

MAURICE B. OTIS

We must learn to recognize the limitations and potentialities of each particular area of the earth, so that we can manipulate it creatively, thereby enhancing present and future human life.

RENE DUBOS

Rivers, streams, and even brooks have played a key role in the development of the United States ever since America was settled. They originally supplied protein from the fish that inhabited them, a source of potable water, a mode of transportation, rich bottomlands for agriculture, and a ready source of power to help launch and sustain developments that have in turn exploited our waters and other natural resources at a rate unmatched by any other country in the world.

As years passed, man returned to these waters human and industrial effluent from the farms, villages, and cities located along the waterways. The presence of these pollutants are readily identified by sight or smell, but the subtle effects produced by nitrogen and phosphorus on the aquatic ecosystems are more difficult to detect. Man also inundated the valleys of hundreds of miles of free-flowing rivers with dams, which in turn formed physical obstructions to previously unhampered movement of anadromous fish. In many instances, these dams created warming of water to temperatures intolerable for desirable fish life.

The future appears no less threatening. Looming on the

horizon are such problems as thermal pollution from atomic-energy and fossil-fuel plants to meet our growing power needs, and introduction of additional chemical pollutants as new commercial products are manufactured for our consumption. Pollution of water by air is the result of particulate fallout in the form of trace metals and DDT. Recent findings in waters of New York's Adirondack mountain area and in Sweden suggest there is a fairly rapid increase in acidity from fallout of airborne contaminants.

The problem of solid waste disposal covering the gamut from garbage to old cars is increasing at an alarming rate. The lowlands, usually sought out for open dumps or landfill operations, provide ready-made leach beds with ultimate seepage of undesirable materials and chemicals to waterways and marshes.

In addition, we have become increasingly concerned with other forms of less subtle changes directly affecting the confines of streams in the guise of "channel improvements." They involve alterations to the streams under such terms as relocation, realignment, straightening, widening, ditching, and gravel removal. These physical changes and the resulting siltation are spelling doom for additional miles of trout streams. They are easily identified by the heaps of gravel bulldozed to the banks of streams and freshly cut or dredged channels.

Documentation of fisheries losses from channelization reveals some startling information. In Montana two trout were taken from thirteen channelized streams for every eleven in unaltered sections of the same streams. There were sixty-nine trout larger than six inches in a section of Flint Creek before channelization, but only six were found the following year. Similar surveys in Idaho on forty-five streams disclosed there were 112 times more pounds of game fish in undisturbed than in channelized streams.

The outcome of silt deposition in streams from channelization and flooding results in covering fish spawning areas, reducing the supporting aquatic food life by cutting off oxygen and light, and filling pools and fish hiding places. Any increase of water flow and velocity in the stream also causes turbidity from dislodged silt particles.

Channelization to contain floodwaters encourages more river valley developments than previously existed. This perpetuates the vicious cycle in a never-ending march called progress, which continues to reduce the number of streams in any watershed originally having water suitable in quantity and quality for trout.

Our flowing-water environments are modified each time man undertakes any sort of development. These changes may range from minuscule to violent, but they always spell tragedy for our dwindling trout waters.

Individuals or groups of concerned persons must exercise constant vigilance on these offenses that chip away at the very source of what many of us consider to be our rightful heritage of game fishing.

The foregoing problems are, in general, of such magnitude and complexity that help and advice should be sought from private conservation-oriented organizations or state and federal fish and wildlife conservation agencies that can provide the expertise to diagnose the problems properly and make firm recommendations for correction or amendment; these groups also can interpret and explain what can be done from a legal standpoint under existing local, state, and federal laws relating to stream protection. There are many federal and state programs aimed directly at providing technical assistance and advice in the area of environmental protection that may also sponsor action programs to alleviate or solve specific issues.

All of the foregoing factors influence the stream system of the watershed. However, of no less consequence is the stream, its immediate banks, and the land bordering the waterway. A practice of new physical treatment combined with enhancement of existing fish-habitat conditions in the stream and landscaping its banks is an important step toward better trout fishing. At least it is if watershed practices and treatments are or will be demonstrating improvement, and stream water quality, flows, and temperature are satisfactory to support trout life. This practice is called stream improvement. Its scope ranges from picking up an empty beer can to major construction of a warmwater fish-barrier dam on the stream at the lowest habitable limit for trout. It offers what may well be one of the best segments of a fish-management program for any particular trout stream because the visual results of such work alone are readily apparent. Restoration or addition of trout habitat, protection from natu-

ral or man-made damages, and the mere creation of beautiful surroundings along a trout stream are in themselves accomplishments worth striving for.

Robert Traver, former supreme court justice for the State of Michigan and well-known author on fishing as well as an ardent practitioner of the art of angling for trout, expressed the trout fishermen's feelings when he wrote, "Trout, unlike men, will not—indeed cannot—live except where beauty dwells, so that any man who would catch a trout finds himself inevitably surrounded by beauty: he can't help himself."

Although the modern trout fisherman has broadened his horizons, and rightfully so, to help solve the larger problems affecting his dwindling source of relaxation, peace of mind, and sport, he can most readily identify himself with his immediate surroundings in and along his favorite stream. Hopefully he will dwell on what can be done to improve his favorite trout stream through preservation and restoration of its natural character and landscape, which are essential to quality angling.

In addition to stocking trout in suitable waters, the legal restrictions imposed on the number allowed for the creel, and protective laws to assure their perpetuation, stream improvement is recommended as a tool basic to the management of trout. It can create conditions that will offer more opportunities for their survival, growth, and natural reproduction. In addition, there is generally an increase in certain food production, which adds impetus to the numbers of trout present.

Stream improvement was first introduced to this country as a stream conservation method on private fishing club streams and the estates of millionaires. Previous to that time it was practiced only on the famous chalk streams in England and Ireland.

In 1930 the Institute for Fisheries Research at the University of Michigan planned and directed some stream-improvement work that was carried out by the Michigan Department of Conservation. This work was conducted on the Little Manistee River in the upper portion of the lower peninsula and on some warmwater streams in southern Michigan.

Fish habitat improvement of warmwater streams was not encouraging, because the response in increased numbers and size of the warmwater game species being managed was insignificant. It is my opinion that warmwater game fish such as large and smallmouth bass are generally underharvested in streams having suitable water quality and quantity. No significant amount of streamside habitat work is necessary to perpetuate and sustain the resident populations except for bank-erosion control in various forms of artificial revetments and the establishment of vegetative stream-bank cover. Such streams are usually lower in gradient than trout streams and have more than an adequate number of pools and deep flat sections to accommodate the existing fish populations. In addition, the cost-benefit ratio of such work does not generally justify the effort, because most warmwater streams are larger in size and therefore require greater works of improvement.

As a result of this initial work and subsequent evaluations, efforts by state and federal agencies have been confined primarily to trout waters. It has been further demonstrated that the best results to date have been on streams suitable chiefly for brook trout, although I believe it is only as a result of less evaluation being undertaken on other salmonid species inhabiting waters where such work has been performed. Resident brook trout population is also far easier to measure than species such as rainbow trout or brown trout, which tend to become migratory as they increase in size and move downstream to larger waters. However, in most cases, stream improvement has contributed significantly to natural reproduction by actual measurement of young-of-the-year trout which, combined with the creation of more pools, will increase the carrying capacity for trout in the stream.

Studies by fisheries biologists of improved trout streams in Minnesota, Michigan, New Mexico, Montana, Wisconsin, Vermont, West Virginia, and New York have revealed convincing evidence of the value of stream improvement. Winter trout survival was as much as 50 percent higher on improved streams, where emphasis was placed on creating additional pools than on comparable unimproved streams. It was also shown that survival during droughts increased significantly.

Installation of various devices and structures on certain streams have shown increases of over 300 percent in the total number of brook trout living in a stream section after improvement, despite a 200 percent increase in anglers and a 360 percent increase in the annual harvest brought about by the additional attractiveness of the area to anglers. In certain cases, catch rates have shown increases from 0.58 to 0.89 per angler trip.

Other studies revealed decreases in silt deposits and subsequent increases in natural spawning areas, decreased summer temperatures that are more tolerable to trout, and decreased bank erosion, not to mention the general physical improvement to the bed, banks, and adjacent bordering streamside areas. In certain instances, extensive stream rehabilitation efforts have meant the difference between a trout stream's death or its survival.

Stream habitat improvement must be preceded by ascertaining that good watershed conditions exist or that vegetative cover plantings are being undertaken to slow surface runoff. Rapid runoff causes erosion and subsequent siltation of streams. Adjacent landowners of all kinds should be encouraged to exercise proper land use through good farm and forest practices. In many instances upland ownership is not confined to farmers alone; suburban dweller ownership is increasing rapidly as farms and woodlots are subdivided to meet expanding housing needs near towns and cities. With this expansion comes improved or new road systems and other developments that, in many instances, also affect our trout waters through relocation or realignment to accommodate these growths. Federal and state conservation agencies are available to provide technical assistance and advice through their staffs of watershed specialists to help bring about the natural harmony necessary between a stream and its watershed.

Where land treatment alone is not adequate to dispel harmful flooding, small watershed developments should be promoted in the form of water-control structures on headwater streams. These structures prevent or retard heavy water runoffs by providing excess water storage in small reservoirs. The outlets of these reservoirs are designed to control outflows which in turn help prevent flooding in the main stream during periods of heavy rainfall. These small impoundments also serve to augment main-stream flows during periods of drought to the extent of their storage and capabilities. In many instances it has been possible to develop a large-capacity reservoir on a small head-water stream where a permanent or conservation-pool level can be maintained of sufficient size and depth to sustain trout. The reservoir's flood storage capacity is built in above the permanent pool level. The United States Department of Agriculture, through their soil conservation service, sponsors many of these fine projects on a cost-share basis with watershed residents, local government, and the state.

The primary objective of stream improvement is the restoration and enhancement of trout habitat. Fishing values are given prime consideration as distinct from erosion and flood control. However, small-watershed flood control, proper land use, and stream-bank erosion control must be recognized as important tools which also should be used wherever necessary to assure success in bringing about ideal living conditions for trout.

Any stream improvement project should be preceded by a survey of actual conditions. This should include investigations of the water chemistry, biological characteristics, and the history of water temperatures, flows, and resident fish populations. Almost every state conservation agency employs fishery biologists who, in most instances, have already made these investigations and can ascertain whether stream improvement is warranted. There is no doubt that many trout streams can benefit from it, but the need is so great that emphasis should be placed on those streams where the most improvement can be realized in relationship to effort. It is certainly more desirable to improve a stream that has the potential for carrying trout on a year-round basis rather than one that only supports stocked trout on a short-term seasonal basis because of critical summer water temperatures. Under no circumstances should improvement work be carried out beyond the capability of the stream to carry trout. Efforts then encounter the law of diminishing returns.

Experience has shown there are many streams where ideal water and temperature conditions exist, but which are almost completely lacking in natural habitat and so have an understandably low carrying capacity. Such a condition is conducive to stream improvement. Another important consideration is the gradient or slope of a stream. A low-gradient or slow-moving

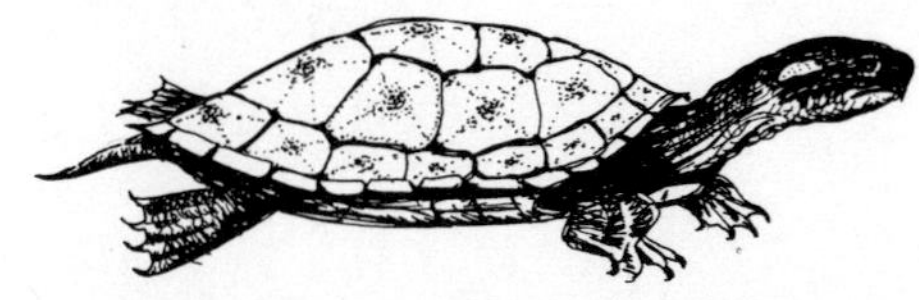

stream is subject to more rapid siltation than relatively high-gradient watercourses. Stream improvement is more easily accomplished in fast-moving streams because advantage can be taken of the water force to work in conjunction with structures placed; chances for reduced siltation over the long haul are also greatly enhanced. However, streams having gradients in excess of three feet of drop per hundred feet in length also have their drawbacks because the conventional, less expensive stream-improvement practices and structures cannot withstand the heavy runoffs and occasional flooding that must be expected.

COURTESY

It would be wonderful, indeed, if everyone lived by the golden rule. This is the first practice the fisherman should undertake when fishing any stream. The trout in the stream belong to everyone, but the land bordering the stream in most cases does not. In many areas of the country, particularly in the East, even the bed of the stream belongs to the upland owner who in all probability loves his land as much as the fisherman loves the stream. The creation of a good relationship between them will form the basis for mutual respect and understanding of problems confronting both individuals. Respect for the landowner's rights will go a long way toward developing an atmosphere of friendliness and receptiveness by the landowner to consider fencing off streams to prevent cattle grazing, leaving green belts of trees along stream banks, tree planting, and stream-improvement structure installations. Every fisherman should secure permission from private landowners to go across their lands to fish. He should park his car where it will not interfere with landowner activities, shut gates, walk along fence or tree lines to avoid planted lands, avoid breaking wire fences, and exercise every courtesy due the landowner as a guest on his land.

Private land ownership, particularly in the eastern part of the country, represents the bulk of our trout fishing opportunity. In New York State, for example, only a little over a thousand miles of the seventeen thousand miles of streams classified as suitable for trout habitation are in public ownership. It therefore behooves not only the individual trout fisherman but also organized groups to bridge the gap of understanding between them and the landowner. Once this objective is accomplished, conversion to a common cause can be more readily undertaken. Stream improvement actually benefits the landowner most of all, because his livelihood, the land, is preserved along the stream, flooding reduced, and erosion of soil along the banks halted.

A simple stream-bank litter cleanup provides a rewarding effort on the part of the fisherman and certainly benefits the landowner who, in many instances, may be grateful to the sportsman for his efforts. On the next fishing trip he can carry along a burlap bag folded up in his creel. It can be wet down to cover fish or better yet to bring home a limit of trash instead of fish. This is a project that can be initiated on his own or he can enlist the aid of others. Properly handled, no landowner can refuse such an offer and it helps pave the way to future projects on the stream in cooperation with the landowners.

STREAMSIDE VEGETATIVE TREATMENTS

Probably the most sound and economical treatment of land bordering the stream is to allow a green belt of trees, shrubs, and grass to grow or to initiate new growth by planting native tree and shrub species. In many instances, growth has been previously cut to obtain the last possible inch of rich soil for crops, destroyed by cattle grazing, or subjected to severe flooding which, in turn, caused washouts of vulnerable bare stream banks and their covering vegetation. Raw stream banks provide huge sources of silt and gravel that can be washed into the stream by high water. The depositing of these materials, as floodwaters recede and water velocities slow down, clogs the stream channel and leaves silt or worse over trout feeding and spawning areas.

Streamside vegetation provides soil stability, slows down floodwaters, and catches silt. The deep, water-seeking roots bind the soil together and the growth above ground provides a concentration of water-flows within the stream channel. Vegetative cover is provided not only to the soil but to trout. Overhanging branches and grasses in the water offer excellent fish hiding places as well as providing them with a ready diet of terrestrial and land-stage aquatic insects that naturally nest, feed, and breed in this vegetative cover. In many sections of stream, the shade afforded by trees along the stream banks provides the marginal decrease in stream temperature necessary to support trout during hot summer days. There are very few streams in this country today where increased water temperatures are needed for trout. Shade from the sun also provides a natural, shifting cover device for trout. Tree plantings along convergent fence lines perpendicular to the stream should also be encouraged where possible. They provide living deflectors to concentrate floodwaters in the stream channel.

Stream banks lacking vegetative cover in sections of stream where normal water velocities do not exceed 0.5 feet per second can be restored by preparing the bank slopes by hand for short breaks in the bank. In the case of long reaches, machinery can be used to shape the banks to tolerable slopes of repose for varying soils to prevent bank slide. Establishment of fast-growing grass types suitable to the soils and successive plantings of shrubs and trees native to the area should be made. Rooted stocks of purple osier willow (*Salix purperea*) are excellent stream-bank plantings. If these are not obtainable, cuttings of black, golden, weeping, or shining willow can be used. Willow cuttings planted in the dormant stage are good but relatively unsuccessful unless used on gravel, sand, and mud bars where they can be planted deep enough to reach low water levels of the stream. The best-growing shrubs generally prove to be silky dogwood (*Cornus amomum*) and red osier dogwood (*Cornus stolonifera*. Plans should include the necessary maintenance and follow-up work necessary to replace grass, trees, and shrubs lost through natural mortality on areas lost by subsequent washouts of weak, freshly planted banks.

Tree planting cannot be termed "instant stream improvement" because the trees and shrubs take many years to mature—often five to ten years. However, it is well worth the effort and time expended, because plantings provide a living stream improvement technique that, left undisturbed, will benefit generations to come.

STREAMSIDE FENCING

In rural dairy areas, the cow is probably the most harmful critter in existence to stream banks. The cow, like motherhood, is here to stay and, like mothers, they can be managed at least

Streamside Fencing and Vegetative Treatment. These three-year-old purple osier willows *(Salix purperea)* were planted in three close rows along each bank and protected from cattle grazing by fencing. A cattle crossing and fence stiles have been provided. Compare the natural vegetative cover between the fenced area and the pasture.

Maturing Streamside Willow Plantings. These well-established purple osier willows *(Salix purperea)* are six years old. In four more years they will reach their maximum height of ten feet. Trees for shade should be planted back of the lines of willow.

Riprap and Grass Planting. Erosion on this stream bank was too extensive for any other treatment except machine sloping and placement of riprap. The ground above the riprap was heavily seeded to establish a luxuriant cover that protects the slope from erosion.

most of the time. Grazing cattle trample banks so that, to all appearances, they look as if a sheepsfoot roller, earth compactor had been towed up and down the banks. Cows graze on the few remaining grass stems and, particularly in early spring, browse on the young green shoots of woody plants and trees to the point that they are killed or stunted and deformed.

Any new bank plantings must be protected from grazing by constructing fence lines along the outer limits of the green belt being established. Normally a fifteen-to-twenty-foot width is acceptable and hopefully, with landowner consent, the area can be widened at points to help eliminate following stream oxbows and sharp bends; these areas are the most vulnerable to flooding and bank washout. Watering and stream-crossing places for cattle or other livestock must be planned to provide at least three or four sites per mile of stream. Select, in cooperation with the landowner, known crossing places with gentle slopes for easy cattle access. There is nothing more exasperating after completing a fencing project than having cattle crossings at the wrong locations. Cattle, like humans, are creatures of habit and will bawl all day for water and never go near an unaccustomed watering place. You not only have an unhappy cow but a disgruntled landowner. Fence stiles should also be placed at fence crossings to accommodate fishermen.

Fencing should not be restricted to revegetated stream banks. There are many areas of browsed vegetation that will flourish once grazing is eliminated. These areas may need supplemental plantings as time progresses to assure ample vegetative cover and shade.

In many instances there are small spring brooks or spring seepage areas that are critical to temperatures and flows of the main stream. These areas should also be considered for protection if subject to extensive grazing. As in the case of the main stream, vegetative cover should be established along their banks but not to the extent that water-flows are reduced as a result of too much water consumption by the bordering shrubs and trees. Small spring tributaries should be cleared of in-stream debris and vegetation to encourage greater flows to the main stream.

STREAMSIDE STRUCTURES

There may be many areas along the stream bank where water velocities and banks are too extreme to stabilize with vegetation without first providing a more firm outer layer of protection from erosion. These eroded bank areas generally occur at sharp bends in the stream and where channel gradients are steep. At first thought, it might seem more economical to eliminate the bend by mechanically straightening the channel. After all, it would reduce erosion and carry away floodwaters faster. Put such thoughts out of your mind, because this is a last-resort effort, and some other landowner downstream must suffer the consequences.

Every stream from its source to wherever it enters the estuarine system has a length set by nature through its own hydraulic system, which remains basically unchanged. If a stream is shortened by man in one place it will lengthen itself someplace else to compensate for the loss in its natural horizontal sinuosity. In addition, increased water velocities caused by de-

Streamside Log Crib. A low eroded bank has been stabilized with stone-filled log crib. Note the streamside fencing and vegetative cover. Large boulders placed in the stream form small pools and trout shelters. The elevated flooring of the crib, submerged to prevent rotting, provides additional fish shelter.

creasing the length of a stream section to a steeper gradient will result in head cutting or bed erosion. This results in deposition of the eroded stream-bed material at some other location in the stream, where water velocities are slower and the stream can no longer wash or carry the material in suspension.

The protection of badly eroded banks can be accomplished by constructing bank revetments consisting of stone riprap, gabion wire baskets or blankets filled with stone, and stone-filled log cribbing. Only these three types of revetments offer any shelter for fish once constructed. The gabions provide the least fish cover. Loose truck-dumped riprap affords excellent cover if the larger stones are jumbled at the base. The flooring of log cribbing to hold the stone ballast provides an excellent fish hiding place beneath the floor. Properly constructed and maintained, log cribbing made from preservative-treated wood or untreated cedar will last for at least twenty years.

Depending on water velocities and local stream fluctuations, it is possible to construct less expensive bank protection and fish-cover devices such as sod- or brush-covered log booms, plain log covers, and tree bank covers. These structures should be used only on streams where water fluctuations do not exceed one to one and a half feet. Adequate anchorage must be assured for these devices because they have a tendency to float when subjected to high water. Placement on sharp bends should be avoided, for if they do float they may be washed up on the stream bank. These structures also afford excellent trout cover and hiding places. Plantings of suitable grasses, shrubs, and trees should be made along the banks behind the structures after they have been installed.

IN-STREAM STRUCTURES

Once bank stabilization to prevent erosion has been achieved, it is time to evaluate the in-stream fish habitat conditions with the help of a fisheries biologist for additional improvement potential. A certain amount of fish shelter is afforded by the streamside improvements. Pools will generally form along riprap and log crib because the force of the water tends to deepen the bed of a stream on the outside of a bend where these protective structures are usually placed.

Pools increase the carrying capacity of a stream because each one will accommodate living space for fish. The pool is the natural resting area for trout; it also provides food in the form of minnows, which tend to concentrate in the pool for the same reasons as trout. Both are afforded shelter from predators by its water depth.

Based on the best information available there should be a 50:50 pool-riffle ratio. Trout rest in pools, but the aquatic food is most plentiful in the riffle areas. It is here that the wet fly and nymph fisherman is at his best, because riffles are prime feeding areas for trout.

With these thoughts in mind, a plan for in-stream structures is made during a stream's low-water period because this is the most critical time for successful trout survival. In the spring, during above-normal flow conditions, many sections of a stream may appear to be excellent for trout, but in low-water periods the same area may be a wide, shallow, flat section where a trout's tail could hardly be covered by the depth of the water. However, it is best to observe the stream under both high- and low-water stages to visualize the effect of the structure planned.

All structures should present a low profile designed for normal summer flows. In addition they should be sloped slightly toward the center of the stream from each bank to provide additional convergence of stream flows during high water. This sloping tends to scour and deepen the center of the stream channel, resulting in a concentration of water during low summer flows in which trout can survive.

With these design criteria, structural improvements can be made that are capable of withstanding floods and ice and do not impair upstream fish migration. When logs are used they are mostly submerged in water, which suppresses decay. At the same time low structures do not appreciably decrease the water-carrying capacity of the stream channel during high-water conditions that might create flooding of adjacent land.

One of the most effective pool-creation devices is a low dam. This is not the ordinary type of dam we visualize, which backs water upstream to form a pool. Preferably, it should be located in a riffle section with an upstream gradient sufficient to preclude little if any impoundment above the dam. Impoundments of any kind in a stream collect undesirable silt and debris. By keeping the dam not over ten to twelve inches in height, siltation potential is practically negligible. The pool most desired for these structures is the one below the dam that is created from the scouring effect on the bed of the stream from the force of the falling water.

Dams are placed in long riffle areas, devoid of pools and having a reasonably hard stream-bottom consistency but one capable of being sufficiently scoured by action of the tumbling water. If soft bottom conditions are encountered, it will be necessary to stabilize the stream bed immediately upstream from the dam by the use of mud sills, planks, wire and sod, heavy stone, or a combination of these methods to prevent undermining the structure. Dams should be located in relatively straight sections of the stream and placed at locations where banks are ample in height and afford stability to prevent end runs of water around the dam. A narrow confined stream section is naturally preferable to a wide stretch because it reduces the size and cost of construction.

Rock Dam. On-site boulders provided the material to construct a series of rock dams along steep-gradient riffle area devoid of pools. The boulders in the pool below the dam serve as trout shelters.

Log and Plank Dam. The force of the tumbling water has dug a fine pool below the dam. The stone-filled, log abutments protect the banks from erosion and concentrate the stream flow. Abutments are built low to offer the least resistance to floodwater.

Log Pool Digger. A single log anchored to a two-log base forms a simple pool digging device. The notch in the top log concentrates low-water flows. The logs extend well into both banks. Stones at this juncture prevent end runs during high water.

Log Crib and Deflector. The deflector directs the stream toward the crib to provide a deep narrow pool along its face. Alluvial material settled out on the downstream side of the deflectors has been stabilized with grass and will be planted with trees. The streamside fencing has already allowed a lush natural growth to develop along the stream banks.

There are many types of pool-digging devices. They range from a simple keyed rock dam or log pool digger anchored into the stream bed and stretching across the stream into each protected bank to a more complex multiple log or stone-filled gabion dam. However, each one provides a concentration of stream flows during low-water periods, aerates the water to increase oxygen, digs a pool and forms a shelter or hiding place for trout under the dam. The pool created is generally of such a size that it is capable of carrying several trout during low water periods. Sometimes, though, a pool is so deep that it attracts a large cannibal trout, which takes over and dominates. It has also been my experience that the water-stilling action of the pool below the dam creates a submerged gravel bar just downstream. Trout then have shelter area, a ready source of food from aquatic life being washed over the dam, and an ideal spawning area close at hand. The last benefit from a dam—and probably the most important—is that a low bed sill is created that prevents stream bed degradation. Degradation or head cutting results in flat, slow-moving sections of stream susceptible to injurious silt deposits.

The stream deflector, jetty, or groin is a device that does exactly what the name implies. Properly located and designed, it deflects water from its location along the stream bank to a point where you want the water to go. It may be used as a single unit to deflect and concentrate the water against the bank opposite to which it is installed. Make certain it is well stabilized from the erosive force caused by the increased water velocity. Deflectors may be used in pairs opposite each other to narrow or constrict stream flows to the center of the stream channel in low-water conditions. A simple rule to remember is that if the cross section of the water area is reduced by 50 percent, the water velocity will be doubled. This rule is applied when designing deflectors to obtain a desired water velocity.

Deflectors used in pairs can be built to create water velocities capable of washing silt off underborne areas of stream-bed gravel suitable for trout spawning. A design to increase velocity in high-water periods deepens the center of the stream by the erosive force of the water, resulting in a deeper stream channel and concentration of water during low-flow periods.

Twin or paired deflctors are, in effect, a low head dam; they create some pooling of water above and between them. They

A Series of Alternating Deflectors. These triangular-shaped deflectors not only narrow the stream channel during low water periods but also increase its horizontal sinuosity. Note the gravel bar formations along the banks.

can be used on alternate banks, bouncing the water back and forth on a wide, shallow, flat, unproductive section of stream. This action, in effect, increases the horizontal sinuosity of a stream and increases low-water velocities that, in turn, keep the stream bed free of silt. The reduced water velocity on the downstream side of deflectors will cause material to settle out, forming gravel and silt bars. Once these bars have sufficiently built up they can be stabilized with grass and willows to prevent further siltation downstream.

Deflectors are also used as channel blockers to cut off low flows in side channels or to direct flows to a more desirable channel. They are sometimes used in series on eroded stream bends to protect new vegetation. Riprap is often placed on the same bank between the deflectors in such a series.

Deflectors can be constructed of logs and should preferably be triangular in shape. They can be built log-crib style, stepped into the bank or formed with single logs, rocks, and rock-filled gabions. Deflectors should be constructed in such a manner that the current cannot undercut them, because they will then lose their effectiveness. Carefully planned, deflectors become an important and useful stream-improvement tool.

SPECIAL STRUCTURES

Fish-barrier dams to block undesirable fish species from moving upstream, fishways or ladders to provide fish passage for migratory fish species, and check dams to trap silt and debris are three other relatively common types of stream-improvement structures. Because of their complexity they should be planned and undertaken only by fisheries biologists working with stream-improvement or hydraulic engineers. The location and construction requires individual designs for each site, based on local stream, soil, and topography conditions. This fact should not deter individuals or organizations who are aware of the need for such structures from seeking advice and help from the proper governmental agencies responsible for such matters. These structures are very expensive to build and maintain. Cost-benefit ratios are a serious consideration in determining implementation of such projects.

Boulder Deflector and Pool Diggers. Boulders added to those already existing form a natural deflector. The current is directed toward the strategically placed boulder pool diggers downstream.

CONSTRUCTION HINTS

Take advantage of natural materials close at hand, such as suitable flood debris, dead trees, reasonably sound old logs, stone, boulders, stumps and even live trees that can be cut with landowner permission. (Live trees for structures should always be cut well back from stream banks in areas where they do not contribute to the well-being of the stream.) Adapt structures to existing favorable stream habitat conditions. Evaluate the results of structures installed to ascertain that they are making a significant contribution to the trout-carrying and holding capacity of the stream. Since experience has demonstrated that structures that work well on one stream may not be successful on another waterway, it is advisable to proceed slowly at first to determine which structures are most successful.

Structures and techniques should be modified to meet conditions. Advantage should be taken of existing situations, such as adding boulders and stones to several already in a position in the stream where a boulder or rock deflector would be desirable. Many times logs washed into the stream can be adjusted and anchored in place to form a simple single log pool digger or deflector. These simple devices, although not as spectacular as a log dam or a log-crib deflector, may be all that is required to meet a desirable habitat requisite.

Avoid artificiality by using natural materials even if the various materials must be transported to the project site. Landscape banks where structures are installed so they blend into the natural surroundings. You may be proud of your work, but there is nothing more satisfying than having a fisherman refuse to believe any stream-improvement work was ever done at a particular stream location.

Structures should be well built and solidly anchored to prevent washouts and erosive end runs around the bank during high-water stream conditions. Build for permanence to avoid having your structures washed out during a flood; stream damage caused by such an occurrence often is more harmful than if nothing had been done. It is more sensible, practical, and satisfying to locate wisely and build a *few* structures well than to lose a lot of devices because of hasty, poor construction.

Unless projects are extensive, simple hand tools, a lot of muscle, sweat and ingenuity are all that is required to undertake a stream-improvement project that will provide equity for the landowner, joy and satisfaction to the fisherman, and a better environment for everyone—including the trout.

Plans for various types of structures, including those constructed from rocks, logs, and various devices using combinations of materials consisting of logs, wire, rock, and plank can be secured from various state conservation agencies, the United States Soil Conservation Service and the United States Forest Service offices. Gabion or stone-filled wire-basket designs are available from the manufacturer. Again, consultation and planning should be coordinated with fisheries biologists and specially trained stream-improvement technicians and engineers.

Stream improvement may be accomplished individually or by seeking help from Boy and Girl Scout troops and 4-H clubs that are looking for meaningful conservation projects to earn advancement awards. Many local chapters of Trout Unlimited have initiated vigorous stream-cleanup campaigns and are actively engaged in stream-improvement projects. Fish and game clubs have formed stream-improvement committees to work with volunteer groups on such projects. Many states have active stream-improvement programs, and volunteer assistance augments their undertakings, for nowadays labor is the most expensive aspect of most stream-improvement projects.

Individual efforts to make a favorite stream a little better each time you fish it also pays big dividends. A classic example of this type of effort has been exemplified many times by Art Flick, famous flytier, fisherman, author, and ardent stream conservationist. I have seen the results of planting thousands of willows, as Art has done or encouraged others to do over the years, on the Westkill, his beloved home stream in the Catskills. He has also frequently been instrumental in initiating major stream-improvement projects and encouraged conservation-oriented sportsmen's groups to do likewise. I have jokingly referred to Art on many occasions as the Johnny Appleseed of the trout-fishing set.

How long does it take a couple of fishing buddies to roll a likely looking boulder into midstream to form a fish shelter or to brace a large flat rock sloped up against the current to form a "tip-up" combination fish shelter and pool digger on the downstream side? Floodwaters may wash these out of position the very next spring but, in the meantime, they will provide an extra living room for trout.

6

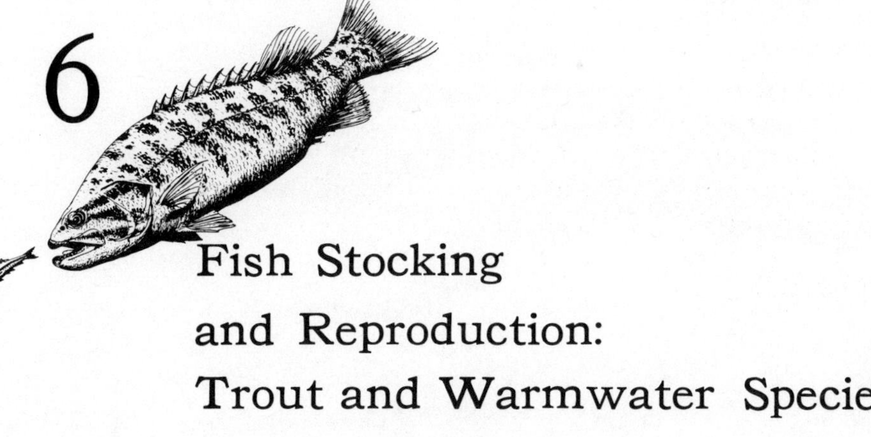

Fish Stocking and Reproduction: Trout and Warmwater Species

DAVE WHITLOCK

"An incredibly delicate and balanced formula is required to achieve a healthy fishery."

On a bright, clear autumn afternoon late last October, I gazed for nearly an hour, hypnotized, at a wild eighteen-inch hen brown trout while she held her position in a moving scene of water and sunrays. I forgot time. I forgot the chill of the water on my bare hands and face, the confining discomforts of a tight rubber wetsuit, the restrictions of the face mask and snorkel to my breathing. What an incredible sight—the long golden-olive form as it turned to the left or right or rose to inspect some object moving along the myriad of floating leaves, twigs, and silver bubbles. Several times the trout pushed her head through the mirrorlike surface curtain to take some insect, perhaps an autumn-chilled beetle or ant. She appeared to fly while gliding through the clear liquid flow of water and sunlight. Her long transparent yellow fins with accented border of black and white reacted like tiny wings to hold or move her black-and-red polka-dotted form in the windlike current. Angled shafts of sunlight lensed by the riffled surface created moving patterns of shadows and light across the trout's back and the stream's gravel bottom. It was a time and sight I shall never allow my memory to lose if I add a hundred to my thirty-eight years.

Spawning salmonids over redd; these are characteristic rather than exact specie.

A few weeks later my trout was joined by a handsome, vividly marked, hook-jawed cock fish. For a few days they lazily circled the pool until nature triggered some instinct by a subtle alteration of water temperature and light. A new urge of life grew within their bodies, some strange new feeling that instilled a new purpose to each, urging them to travel upstream through many pools and riffles. The hen fish chose a certain area with the right texture of gravel and a perfect combination of water flow and depth. Here she soon began to build a redd, or nest, that would catch and hold her eggs for incubation. She lifted the coarse gravel with violent sweeps of her tail; the current then quickly washed the turned stones free of fine sand and sediment particles. A shallow depression soon formed in the stream bed and directly behind it a low mound of sized clean gravel. A clear springlike flow of warmer water seeped up through the redd's bottom, and the hen fish sensed this would be ideal for her spawn. As she worked continually to build her redd, the cock fish established the area as his territory by chasing away all other fish that would later be a threat to the eggs as they were deposited and fertilized. Several times he tried to help with digging the redd but with only limited effort.

The pair of trout now settled side by side in a newly created depression, and the hen fish began to lay her eggs with rhythmic body action. As each group of a few dozen eggs dropped from her vent, the cock fish with similar movements fertilized them with a white cloud of milt. The bright reddish-orange eggs separated and sank immediately into the lower end of the depression. They were moved in a rolling action by the current and the trout's tailwash into the many tiny crevices created by the coarse gravel mounded at the rear of the nest.

As each egg settled into a depression it gradually moved deeper into the loose gravel until it rested hidden near the bed of the stream. The egg's outer membrane allowed the water to seep into it and soon it swelled to nearly twice its original size. The outer membrane then toughened to form a rubberlike protective shell. Here each egg would lie in semidarkness for many days until the incubation period was accomplished. Each day the little trout embryo, curved within the egg, grew a little and then began to wiggle around within the shell. Soon the delicate egg-sac-laden trout fry would break through a certain weak spot in the shell and wiggle free, leaving the empty shell. Now it would begin a new phase of its natural life struggle for survival in the stream.

The pair of fish remained several days at their redd, depositing and fertilizing each group of eggs. After each few hundred eggs were laid, the hen would dig more gravel to create new incubation areas for the next set. She laid only three or four thousand eggs this fall, her first spawn. Later, as she grows older and larger, her ovaries will provide twice that many.

When the last of her eggs were laid, she found enough strength in her tired body to work the gravel back into the depression so that the ever-present erosion of the current would not destroy the redd.

Not all eggs hatched at the same time. Those that lay deeper in the redd were first because the warmer seepages from beneath the stream bed caused faster development. As the egg sacs laden with fry worked between the coarse gravels, they soon absorbed their clumsy orange stomachs; as the yoke was digested into energy and growing body parts, the little fry looked more and more like a real little fish.

The days arrived when each fry no longer had any yoke to feed on. Now they were darting and swimming through the cavelike channels of the redd, and they began to strike at various bits of food. They found it exciting fun and were soon feeding on almost anything that wasn't as large as themselves.

With their new ability to move and swim, they by instinct began working up toward the top of the stream bed. Finally each one wiggled through the last few stones and out into the free-flowing current to begin life as a wild free trout. But there were only a few hundred fry left from the original three thousand eggs. Each day and night their numbers dwindled as aquatic insects and crayfish mined the gravels for the helpless little trout. As they worked near the top, some were caught and eaten by the stone catfish and sculpins that lived under the loose rocks. These few hundred would be further decimated in the fight for life, until only a dozen or so would mate and reproduce in the third year of their life.

Both the hen and cock fish bodies showed discoloration, weight loss, strain, and wear from the exertion of spawning. The hen now turned wearily away and slowly rode the current back to the place where she began her spawning run. The cock fish remained at the redd site for several days, driving off other fish and looking for another hen without a mate. Then he too, thin and dark, weak from the lack of food and the rigors of spawning, rode the current to his old lie, where he would rest his spent body and break his month-long fast. A few weeks later their bodies began to fill out again. The bruises and split fins healed as the metallic luster came back to their skins. As they both slowly returned to their prime condition, their third year of life was completed. Upstream, their spawn was hatching and these new fry were just beginning their first year. The cycle was complete.

I, like a lot of other fishermen, thought I had a good knowledge of a fish's life cycle from studying college textbooks, reading all sorts of articles, seeing nature movies, and from casual personal observation on days spent astream with rod in hand. I was satisfied I understood the whole scheme and even proudly spoke and wrote of my knowledge, passing it on to others for their enlightenment. Then with such pseudo expertise I had the opportunity to make a boyhood dream come true by initiating a unique program for Green Country Flyfishers of Oklahoma that involved the creation of a new trout stream. Two years of initial stream study were spent to determine if the stream conditions were right to support trout all year. An involved study and

search were made for the right stocking method, and then finally a three-year application of our stocking plans established a successful wild, natural, reproducing brown-trout fishery. This experience and research finally opened my eyes and mind to a totally new world. How utterly amazing and incredible is a stream fitted with the harmony of all life forms.

I doubt if any fisherman or naturalist would argue with the statement that "a fish is the most important form of life in any stream." Yet we take so much for granted about its existence there. A stream-bred adult trout or smallmouth bass is a priceless treasure, living free and giving water life, wealth, and promise beyond our comprehension. Each such fish represents the best that nature has. Such a specimen of any particular species is the distillation of one from ten thousand or more that were rejected by the cruelly honest and objective law of natural selection.

A stream devoid of fish life is considered "dead water" and holds little more attraction than a concrete drainage ditch for the fisherman. However, today we are faced with ever more "dead water" situations, as civilization digests our land, air, and water into lifeless excrement that even the crudest life forms are hard pressed to survive on. No longer can we take for granted that water flowing over a bed of rocks, gravel, sand, and sediment will spontaneously produce fish and other life forms for them to feed on. An incredibly delicate and balanced formula is required to achieve a healthy fishery; abnormal changes in temperature, pH factors, water levels, chemical composition, the introduction of new life forms, or excessive removal of native life can have drastic and devastating effects.

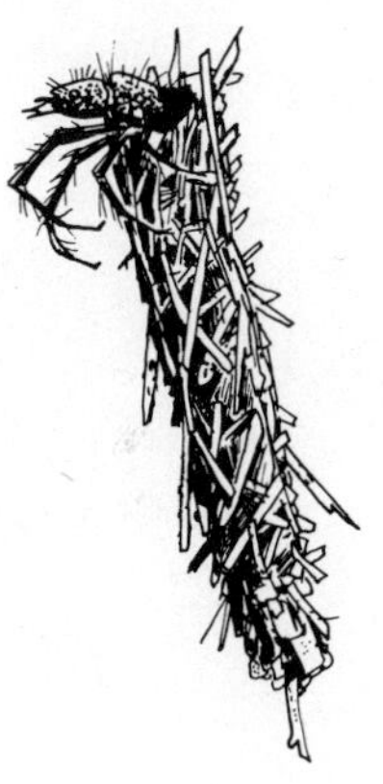

To keep a river or stream populated with native or exotic fish to provide food or sport we have three basic alternatives.

1. Natural spawning by the mature resident or migratory fish to populate the waters;
2. placing in the stream eggs that have been artificially spawned;
3. hatching and raising artificially spawned fish to a desirable size for live releases into the stream.

Each method works well if it is intelligently used; and laws, restrictions, and seasons are additional assistants to protect stocks. The final verdict always lies in the hands of Mother Nature.

JUNGLE BENEATH THE SURFACE

I have spent countless hours beneath the surface of many streams, quietly observing and photographing life there. As I began to see and feel the real environment that exists there, I soon realized there is much more than can be seen from shore or while wading or even in the descriptions in textbooks. Imagine the most dangerous jungle on earth, full of wild animals and having weather conditions with great variance and violence. This in effect is the fish's environment in the stream. Day and night the dangers never let up. There is no escape from the constant threat of vicious predators. A trout, from the minute it enters the stream as an egg or fish, will be threatened with extinction from every conceivable type of monster. Aquatic insects, external and internal parasites, eels, lampreys, crayfish, disease, fungus, other fish—all are a twenty-four-hour-a-day danger. Trout have a constant need for protection in this nightmare meat-grinder environment that renders life a poor chance. From above come snakes, birds of prey, mink, otter, raccoons, and man, potentially the ultimate killer that bends or breaks the whole balance of fish population.

Not only must a trout contend with being killed by predators, it must go directly among them to compete for food to nourish itself for growth and strength. Often food is scarce due to competition and seasonal changes.

Fish, being cold-blooded, are influenced greatly by temperature. Most fish have a range of 20° to 30° F, which allows the optimum physical and mental ability. Extremes of colder or warmer water sharply limit their metabolism—which governs all the life functions, digestion, growth, energy and total mental ability; all these functions go into slow motion during frigid winter lows or baking summer highs. Warm-blooded animals are affected to a much lesser degree.

Streams have storms many times more severe and lasting than the worst killer storms of air and land. Land storms producing temperature extremes and heavy precipitation affect streams unmercifully, for streams eventually absorb all the compounded effects of such a storm. As rain runs off the land, it creates a

water storm within the area, concentrating itself viciously on the streams. Such water storms may last for hours, days, even weeks. The intensity of the flood tears great amounts of the land and bottom out, creating choking sediments that displace oxygen and damage the delicate fish gills. As fish try to escape the fury, they often fall victims to land predators or become land-trapped by rapid changes of water levels. Such storms are especially cruel during winter months when fish are practically helpless from the numbing cold, water's force, and floating flood ice that tears and rips the heart out of the stream. After the storm subsides, fish must adjust to the fast-falling current, and clearing water often reveals strange areas full of new hazards. They must then relocate themselves and establish a new daily routine.

Drought brings another threatening time for fish, for their world shrinks and the water usually becomes less and less livable. Protective depths are scarce, and movement any distance up- or downstream may become impossible. Summer drought creates excessively high water temperatures, winter drought brings killing freeze-ups, anchor ice, and critical oxygen shortages. Without suitable inflow to keep the water purified, it becomes choked with organisms that infect the fish and cause a slow wasting death. Such stagnant water is void of the necessary dissolved oxygen, and thus the fish choke and die in it.

Day and night, week after week, all year long, animal life in any natural stream can never relax. It is a struggle almost beyond our comprehension. No jungle on earth holds more dangers for its inhabitants than a natural stream. A fish has no help other than its strength and instincts; there are no friends, no protective laws or policemen, no doctors or medical aids, and most of all no opportunity to make up for mistakes. Death is usually the sentence for weakness or error.

When we undertake the task of stocking or restocking a stream, methods must be used that give the fish a chance to adjust or adapt to the environmental challenges of stream life. Just dumping live fish into the stream without planning can be extremely wasteful, expensive, and foolish. The very best approach usually results in large losses if not total mortality. Probably 75 perecnt of all attempts to establish better stream fishing—especially for trout, steelhead, and salmon—fail to some degree because a reasonable evaluation of the obstacles facing the stocks has not been made. The Green Country Flyfishers' work, and other projects I worked with, have been marred because we lacked understanding of the associated problems. Introducing disoriented fish into wild predator-oriented water and not correctly evaluating the environmental extremes are the two main reasons most such attempts fail.

Wherever possible it is far more practical to encourage the established native species to increase their population *naturally*. This is accomplished by improving living conditions, eliminating excessive population predators or competing species, and creating enforced protective fishing regulations—and also providing more areas, or improving existing ones, that encourage more successful natural spawning. (Many of these methods are discussed in detail in other chapters of this guide.)

Resident native species possess the unique combination of abilities needed to cope with a particular stream. No two streams are exactly alike, so the native resident is just as unique as its home waters are. Exotics or hatchery-domesticated native stocks have all the natural hazards facing them, along with this added handicap of adjustment. Often they are not capable of this adjustment.

PUT-AND-TAKE METHOD

Put-and-take stocking is a popular method presently employed by most states and federal trout hatcheries. Brood trout are usually rainbow or rainbow hybrids, grown and held in confined ponds or raceways. A few hatcheries use limited amounts of brooks and brown trout. These brood stocks are the results of careful fish culture that has produced several strains of fish that exhibit the following traits:

1. high resistance to disease
2. rapid growth uniformly and weight gains
3. low cannibalistic tendencies
4. good tolerance to temperature extremes
5. domestication
6. high reproduction rate

Both brook and brown trout, though quite popular with most anglers, have various traits that cause them to be impractical to raise in captivity or for stocking most streams under put-and-take management.

Mature ripe fish are stripped of their eggs and milt; then fertilization and incubation are managed by fish culturists, usually in large indoor facilities. Once new fry are able to feed and begin growth, they are transferred into efficient concrete raceways or similar constructions supplied with a good source of cold moving water. Here they are fed and cared for twenty-four hours a day by attendants. Problems, from disease to cannibalism, are quickly and efficiently taken care of; as the trout grow, they are divided and redivided into additional raceways to accommodate their sizes, carefully segregated as to size and

Modern trout hatchery raceways. Note the autofeeder at the right. *Nelson Renick*

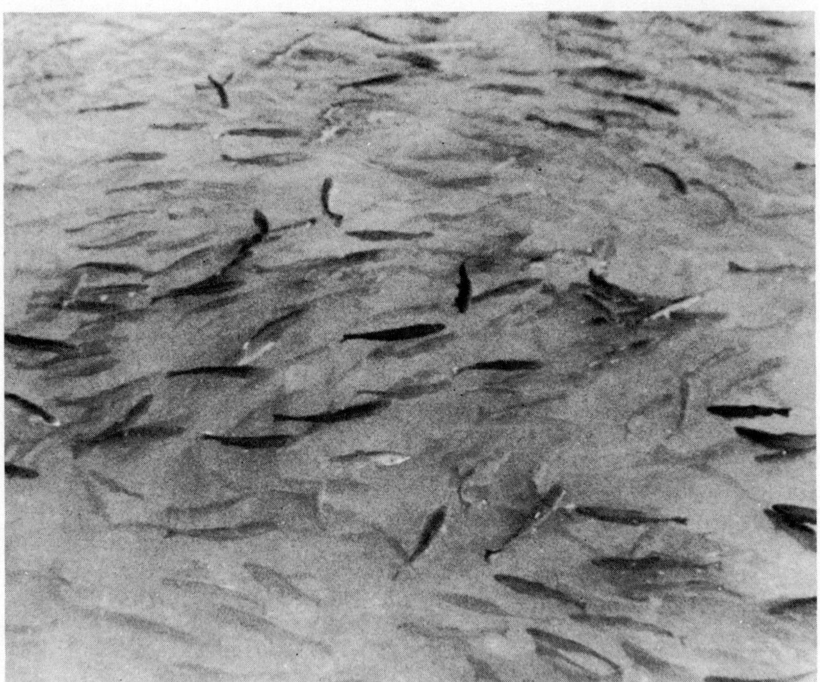

Trout in a holding area, or raceway, readying for stocking. *Nelson Renick*

age. Water sources having ideal and constant temperature ranges are highly desirable for rapid year-round growth. Usually large springs or impounded cold-water lakes make the best water supplies.

The holding pens used for raising trout are usually long narrow affairs with aerated running water sources at one end and a drain at the opposite end. This design provides each trout exercise but doesn't consume much energy, which would discourage weight gains or require more feeding. Most of these raceways are made of molded concrete. I have seen such holding areas formed by merely digging them in areas adjacent to water sources and lining them with gravel. Though these natural raceways allow more realistic environments, they, like small ponds, create certain cleaning and maintenance problems.

One of the most recent developments for retaining growing stocks are the cylindrical raceways. A silolike column stands upright, filled with water and fish. Circulation of water goes into the top and out the bottom, carrying food and oxygen to all fish within the raceway. Gravity and current flow keep the silos flushed of waste products. These raceways are so designed that they save much space—and they are reputed to produce very fit fish. Another is the development of large floating pens. These floating raceways are used primarily in lakes that have good water conditions through periods long enough to have fish reach release sizes. Floating food pellets and very low maintenance cost make this new method extremely practical. Unlimited space, some natural food, and near-natural surroundings improve fish condition and coloration somewhat. A number of such programs are in use where lakes are not frozen over through the winter months. Arkansas has such programs now in use. With unlimited amounts of quality lake water available, that state has found ideal conditions for the floating raceway pens from fall through late spring, giving ample time to grow trout to release sizes.

Success of the put-and-take program is usually gauged by three factors:

1. the production of keeping-size fish (usually 8″ to 12″) in the shortest amount of time
2. a high percentage of catch of stocked fish within the shortest period of time (usually 70 to 80 percent is considered good)
3. a high angler-success ratio, without regard to angling methods or sport involved

Since these put-and-take trout are primarily raised for stocks to harvest by anglers immediately, there are several strong objections. First, in most such hatchery programs these fish are totally *domesticated.* They are more pets than wild creatures. Man has cared for their every need and provided total protection from 95 percent of the harmful elements. Breeding of many generations of domesticated hybrids has all but wiped out natural instincts. Trout that were once strong in their heredity have lost their best traits through a selective breeding that prized hatchery manageables highest.

Such fish, when removed from their home raceway or pond, hauled to strange wild water via truck and dumped into the water, undergo considerable shock. They are tame and helpless and confused. Only their size and body fats allow them to survive for long. Soon the absence of regular free food handouts cause the trout to undergo panic hunger pains. Unable to understand their plight, they struggle to stay schooled. Any object of reasonable size is swallowed without caution or fear. Podded together, scared and starving, most of them fall victim to the angler or other predators. When these trout are hooked, they are confused at their sudden inability to move freely—and they usually struggle without much coordinated effort. This fight lacks the directional runs or clean jumps of a wild trout. A wild fish immediately realizes he is in trouble and uses his wild strength and instincts to rid himself of the hook.

Several years ago, while attending a fly-fishing conclave, I had an awakening experience regarding hatchery trout. A pool near the main meeting area was stocked by a local trout club with about a hundred fat 10-to-14-inch rainbows, furnished by a state hatchery. There was to be a fishing contest for women and children the next day, and some of these fish were tagged with numbers that represented nice prizes.

I watched the trout circle and dart around the small cold clear pool that afternoon. Even these hatchery trout possessed an enchanting beauty that tingled my imagination with visions of past days astream, fighting leaping pink-and-silver bows. I picked up several small pebbles and a matchstick from the walkway and began to pitch them, a bit at a time, into the pool. Each foodless morsel was immediately encircled by a rush of spotted forms. Suddenly the trout all congregated there at my feet, begging for food. I squatted down and touched my finger to the water's surface. Each time the tip entered the water, it was immediately nipped or nibbled by the begging fish.

All I could relate this pitiful sight to was a flock of fenced-in, tame white chickens I had hand-fed near my home a month or so before. Where could there be sport in such a thing? Catching such pets would be the same as shooting the affectionate food-begging chickens. Most put-and-take trout differ little from such chickens.

In defense of put-and-take trout stocking I must concede that the method does provide angling for thousands of us. This is especially true in areas that could never normally support wild populations of trout under high public angling pressures or environmental extremes. Some well-managed hatcheries, private and public, produce a well-conditioned fish with at least some good qualities of wild fish; they may even be distributed intelligently along the length of a stream, rather than dumped. Such efforts are more costly, requiring more manpower, water space, and supplemental natural food diets in combination with the prepared commercial foods. These fish are usually in better physical condition, colored more like wild fish, and do not carry the body disfigurations resulting from overcrowded conditions.

The put-and-take method of artificial stream stocking is a mixed blessing. While it usually provides the angler with live fish to catch, the fish are of limited sporting quality and seldom are able to adapt, spawn naturally, or attain much size.

PUT, GROW, AND TAKE

The put, grow, and take stocking method is quite similar to the put-and-take plan except that when the trout are released into a stream or lake they are expected to adapt and grow before they are caught by anglers. Fish used for this method are usually released when they are much smaller than in the put-and-take program. This plan is usually used where streams, rivers, or lakes are rich in natural foods and have restricted seasons or extensive areas that provide good sanctuary from fishing pressure. Salmon, steelhead, and migratory strains of brown and brook trout are often stocked by this method, though there are many nonmigrating put, grow, and take programs across the country today that are very successful. The brown trout is best suited for such a program.

Basic requirements for put, grow, and take fisheries vary from area to area, though a few remain common to most waters. First, fish released must be large enough to discourage most existing predators. Pretreatment of waters to remove excessive predators or competitive species implements this initial requirement, and poisoning, netting, electric and chemical treatments are means of controlling this problem.

Second, waters must provide a good year-round environment for sustained life and growth. Seasonal extremes can cook or freeze-dry or drown stocks. Lack of abundant food can cause stunting and imbalances and eventual fishery failures. Abundant natural food provides free dining and fast growth, reducing high management costs.

Third, if fish are to develop to expected proportions, there must be either strict regulations to protect them for premature removal or restricted or inaccessible areas where a majority of stocks can find sanctuary. Streams that flow directly or indirectly into lakes or oceans are usually best for a put, grow, and take fishery; these larger areas provide stocks ample sanctuary from weather, predators, and angling extremes.

Such a stocking program has many desirable traits. If a stocked fish adapts to a wild environment, its physical and mental condition changes drastically. The angling potential such fish offer approaches that of their wild cousins, and the longer these fish remain free the more promise they provide for real angling challenges. As a table food they rank extremely high over hatchery-fed fish and rival or excel their wild counterparts.

Since most put, grow, and take programs acquire their brood eggs from trapping wild fish during their spawning runs and releasing the young fish much sooner, considerable time and expense are saved. These savings help extend good management to other important areas needed to provide a total program. A final bonus to this method is the excellent change of "put" fish reaching maturity and accomplishing a natural spawn; in most put, grow, and take fishery-management areas, I witnessed some degree of natural spawning occurring. The most unusual example I've ever seen of this occurs each fall in the tailwaters of the White River below the Norfolk and Bull Shoals dams in northeastern Arkansas. Each area has a fair number of stocked carryover fish that mature and mate to spawn. Water levels fluctuate up and down daily, from inches to several feet, because

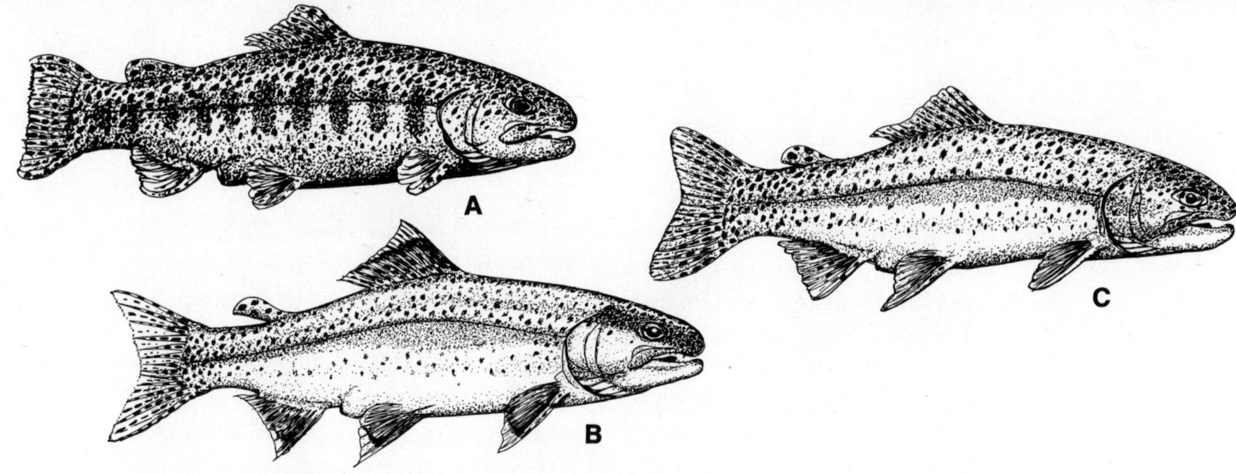

Diagram comparison of three stream-living ten-inch rainbow trout, identifying each trout's background.

A. Freshly released hatchery fish "put-and-take plan"—note characteristics:
   1. general overall dull dark coloring, especially sides and underparts.
   2. fins, especially dorsal and pectoral, badly scarred and stubby. Tail is shorter, thick, and generally small compared to body size. Adipose fin is thick and large.
   3. Parr marking usually apparent due to young age despite size. (Because of forced high feeding and rapid growth.)
   4. Vent, swollen and enlarged due to large volume of food consumed daily.
   5. Body short and fat, tail wrist thick from lack of hard exercise.
   6. Spotting very heavy because of proximity with other dark objects (fish and raceway walls).

B. Natural-spawned or Vibert Box trout.
   1. General overall silvery look, especially on sides and back. Stomach is very light colored.
   2. Fins very long and have transparent feathery look. Usually will be tipped with white, accented with black and crimson backgrounds. Dorsal fin and tail almost appear to be enlarged.
   3. Body very streamlined looking despite body weight due to superb physical condition of stream exercise and natural rich foods.
   4. Spotting fairly heavy on tail and dorsal fin, but body shows light spotting because of light backgrounding in clear water.
   5. Mouth will have well-developed teeth; vent small thanks to high exercise ratio to daily food consumption.

C. Carryover or early stocking fish (put, grow and take) program.
   1. General overall coloring good, vivid, and often silvery like Trout B.
   2. Some dullness on sides and stomach if trout is season's carryover. Fins, especially dorsal, caudal, and pectoral, show slight to bad scars or shorting. Usually lack the transparency of wild fish, but better colored than newly stocked fish.
   3. Probably no parr marking, or very faint. Lateral stripe will be more vivid than either A or B.
   4. Spotting will be fairly heavy on sides and fins but metallic sheen is beginning to mask lateral ones as in wild fish.
   5. Carryover will usually be thinner than hatchery fish, but will not appear in as good physical shape as natural wild fish.

of the hydroelectric and flood-control programs there. Fishery biologists have said that under such circumstances little or no spawning occurs successfully. Though they are in part correct, I have witnessed a most amazing adaptation to this situation. I'm positive these few fish do spawn successfully and that some fry are produced. Here's how the fish work it.

Few species of fish spawn if water levels are unstable, especially the nest builders. I doubt if many wild salmon or trout ever attempt to spawn during flood conditions. But these White River trout have adapted to the extremes and irregular water fluctuations in their day-to-day existence; each older fish learns the "feel" of where the water will be most stable and this is always of course at low tide when only minimum flow is present. With this same keenly developed instinct, they choose a spawning area that has ideal depth at low tide and good stable gravel bottoms. Then, only when low tide is in effect, do they actively spawn. As soon as the tide begins to run higher they quit their activity until conditions are again stable with low tide. I have never in my sixteen falltime periods on the river found a redd above the low-water mark, even during periods of prolonged high water.

There are a few objections or disadvantages to the put, grow, and take programs. Most require more time to provide sizable fish. Fewer stocked fish survive unless the water is practically free of natural predators—or an extremely perfect environment; and a long lapse of time will often find fish moving out of the stocked areas, decreasing angling success ratios. But none of these objections holds much significance when compared to the advantages of this method.

NATURAL SPAWNING METHODS

Fish that exist in a stream from birth, or develop their maturity after an early stocking such as I discussed for the put, grow and take method, are good prospects for natural reproduction. If conditions are normal it can be reasonably certain that spawning will occur. In fact, there are still many freshwater species that have never been successfully spawned in a foreign environment such as man-operated hatchery programs: smallmouth bass, muskie, and flathead catfish—highly desirable game and food fish—are extremely difficult to propagate, requiring

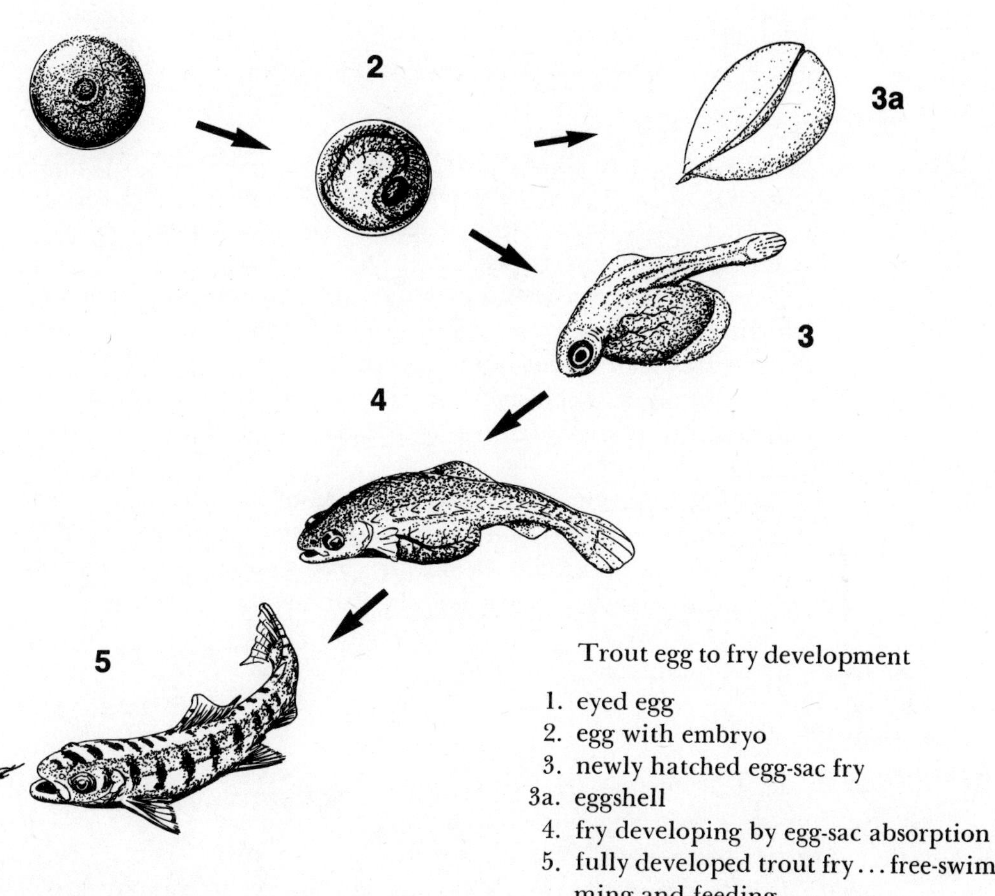

Trout egg to fry development

1. eyed egg
2. egg with embryo
3. newly hatched egg-sac fry

3a. eggshell

4. fry developing by egg-sac absorption
5. fully developed trout fry . . . free-swimming and feeding

complicated care and feeding. In such species, natural spawning is the only feasible method to insure adequate populations for angling or food uses. Therefore, man's role must be indirect, providing protective measures to insure the survival of enough adult fish to carry out natural reproduction. We also can assist by preventing damage to the stream and making improvements that will provide better areas for actual spawning and fry survival.

Earlier I described the basic spawning method of the brown trout. Most species of fish living in streams follow similar methods. A few others scatter their spawn over large areas, and the eggs of others float free until they hatch. But all have one thing in common: *almost unbelievable mortality*. During incubation most of the eggs are destroyed or eaten; usually less than 15 percent survive long enough to hatch. The young fry are constantly attacked by a myriad of predators. And so it goes until a mere few survive long enough to spawn themselves.

However wasteful or cruel nature seems to be to her creatures, her method insures perfection—the perfection of each generation that will insure a subsequent generation capable of withstanding natural hazards of life.

Many of us fear that tampering with this unique scheme by allowing wild stocks of fish to be interbred with hatchery degenerates will eventually destroy the quality of our wild stocks and render each such species lame against the demands of nature's severe law. This would mean the extinction of certain species.

Natural reproduction and population maintenance seems more easily accomplished with most warmwater species than with cold-water species: warmwater fish seem to produce more eggs and spawn at closer intervals. But both are faced with the

Left to right: Bob Cunningham, Bill Greenway, and Roland Been, planting brown-trout eggs in a spring creek in Oklahoma as a Green Country Flyfishers team. *Dave Whitlock*

common problems of stream life and reproduction. A stream will support only a given amount of life; it has a higher ratio of shoreline to water surface area and a very low depth to surface ratio. These facts expose stream life to many more varied elements than other bodies of water, and fish populations and their natural reproduction will always hang in fragile balances in a stream.

I recommend the following steps to encourage the best chances for natural stocking by the spawning wild native fish:

1. Establish a closed season during spawning periods.
2. Limit the killing of adult fish by:
   a. restricting the number of fish taken
   b. setting maximum size limit
   c. regulating angling methods strictly
   d. fishing only with one-hook artificial lures
3. Provide good watershed conservation methods to eliminate erosion, pollution, sedimentation, and senseless waste of water.
4. Educate ourselves and our fellow anglers to the true meaning and value of angling for a wild fish and the significance of its release.
5. Join and support the Federation of Fly Fishermen, Trout Unlimited, and any other local active outdoor organization.

THE VIBERT BOX SYSTEM

The Vibert box, a uniquely designed trout or salmon egg incubator, offers a new practical alternative to the dilemma of ineffective and expensive popular stocking methods. Dr. Richard C. E. Vibert, a Frenchman who has spent his entire adult life researching fishery problems for his country, designed his box and method in 1949. Dr. Vibert had discovered from his research and observations that modern restocking methods were falling far short of their expectations to maintain high qualities of trout and salmon stocks in his country and elsewhere. He devised his hatching box to take advantage of the most important principles of nature's methods.

The Vibert box can be filled with up to 1,000 fertilized trout or salmon eggs and placed or "planted" in the same general area

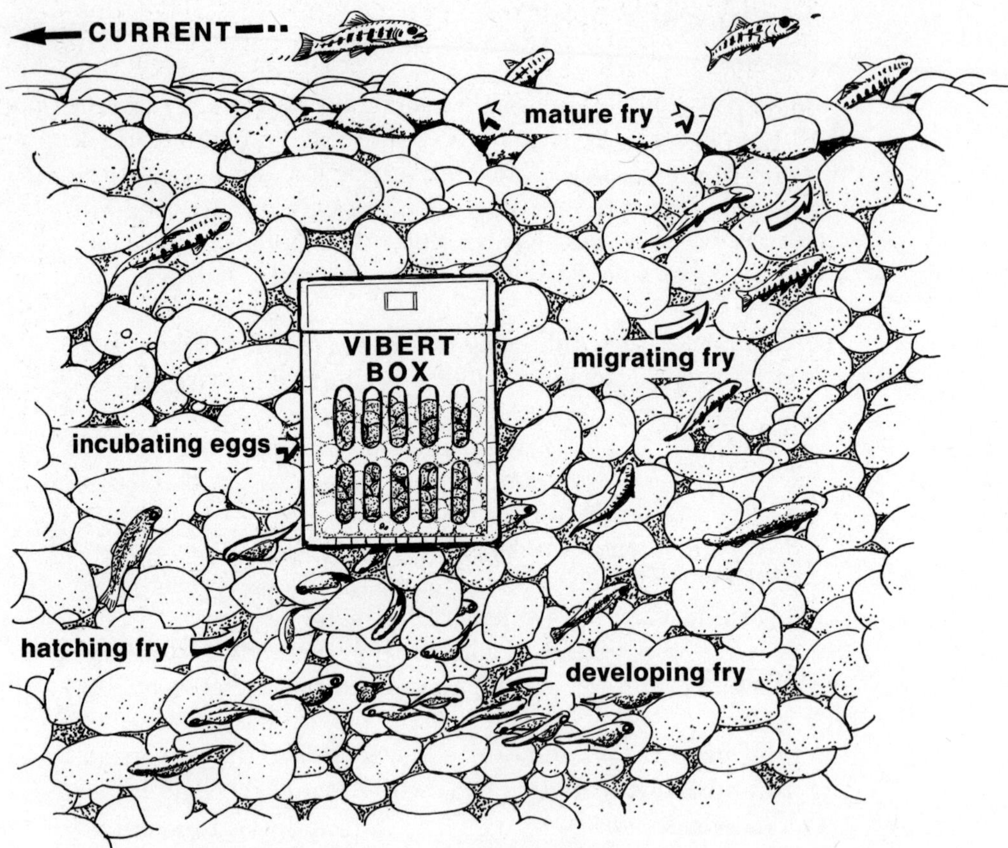

Stream-planted Vibert Box. Beneath gravel horizonal view *(cut away)* showing all stages of fry development.

of a stream that like species would normally use to spawn. This molded clear plastic box is slotted on all sides. These slots are ⅛″ × ½″ and serve to allow water to flow freely through the box but restrict eggs from loss. As the eggs hatch and the fry become active they can pass easily out of the open slots. The basic mechanism of this method is extremely simple yet nearly perfect in accomplishing its purposes. This simple box allows the natural incubation of eggs within the environment as wild fish are produced, yet it creates a protective sanctuary against the dangers that threaten naturally spawned eggs and newborn fry.

When a pair of trout or salmon spawn, their eggs are incubated in the redd, but unlike some species—such as bass, which

Three views of the Vibert Box—with lid off, with stream attachments of anchor and locator line, and with a size perspective in my hand. *Nelson Renick*

A Vibert Box with five hundred brown-trout eggs incubating beneath the gravel of a spring creek in Oklahoma. *Nelson Renick*

guard the nest from predators until the eggs hatch—the salmonids do not protect their eggs. While only 15 percent of eggs naturally spawned will hatch, with the Vibert box the same eggs in the same environment often produce hatches of 90 percent or more. Not only does the Vibert box allow up to eight or nine times more fry to be born, but there is positive evidence from Dr. Vibert and my own work with the system that the method produces stronger and more perfectly developed fry than either natural stream incubation or artificial hatchery methods do. When the Vibert box is properly employed it allows optimum potential of healthy stream-hatched fry development from either domestic or wild stocked eggs. The Vibert box allows the egg to take full advantage of the ideal natural elements found only beneath the gravels of healthy streams. These are usually absent in an artificial hatchery. It also gives the egg protection from the harsh elements of that same stream. This combination of ideal effects and a limitation of irritating or harmful effects allows the delicate embryo to develop more perfectly.

By hatching any salmonid egg, regardless of its parents' quality, in a good natural environment the fry are immediately subjected to the demands of their stream; and their early adjustment is far more efficient than the process of introducing older fish. Those born there, and that pass the test of nature, always show stronger traits than older hatchery stocks of the same parents, displaying far more ability to cope with environmental changes and hazards.

In most public waters these days, *ample* natural reproduction is just a dream from the past and escapes practicability in all but the most guarded private or managed streams. Yet by using the Vibert box system we can either greatly increase the percentage of wild-trout-spawn success or stock the waters with the best domestic stocks. The Vibert box is not expensive to use, nor is it time consuming. It especially lends itself to groups such as sportsmen's clubs that want to improve a local trout stream. In 1970, when Green Country Flyfishers made their initial Vibert box stocking involving some fifty thousand brown trout eggs, it cost them less than $300. Manpower was donated by the club members; twelve members planted the fifty thousand eggs in a hundred Vibert boxes on one Saturday. Boxes and eggs are a little more expensive now but not much more. It would have cost at least ten times as much to grow or purchase enough quality trout to stock the stream with a comfortable number of established adapted fish. No trout we could have purchased at *any* cost would have come close to the Vibert-box-stocked browns that now thrive and spawn there. This is the crucial point: to understand that the fish that inhabit any natural stream are only worth as much as the method that put them there. Strangely enough, the "best" cost less, and each degree below the optimum we go the more it costs us in time, money, and sport!

The Vibert box has been used very successfully in several European countries and here in the United States by fishing clubs representing Trout Unlimited and the Federation of Fly Fishermen; some state fishery departments are also experimenting with the method. Until recently, however, it has not received interest commensurate with its promising potentials. This was due to two factors. First, no extensive research program was undertaken by a responsible agency to develop data and publicize the method. My work with the Green Country Flyfishers provided the ideal field testing project, and my observations and results plus my additional test programs sponsored by the Federation of Fly Fishermen have given us the basic material to develop this material nationally.

Vibert Box slot function. Salmonid eggs will not pass through ovular slots, but slot shape permits water circulation and frees young fry easily. Slot also prevents most predators from reaching incubating eggs.

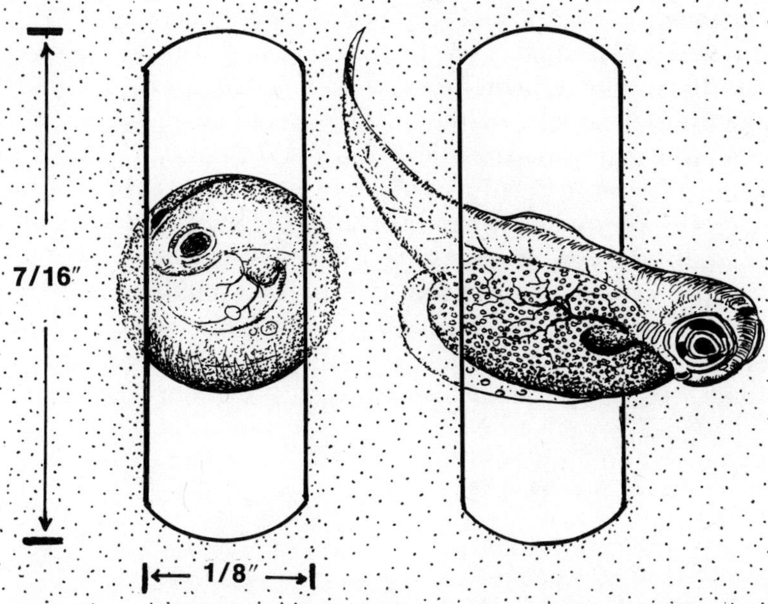

Today, with most state and governmental agencies chiefly oriented to large put-and-take or put, grow, and take restocking methods, the Vibert system does not enjoy much official enthusiasm. This is mainly because it takes at least eighteen months on a national average to produce catchable trout with the Vibert box or natural spawning method. Quick fish crops and a high harvest—the goals of those who seek tourist dollars—can't be accomplished in the Vibert box manner.

Some states are now exploring the Vibert box method to maintain or increase their dwindling wild natural stocks, especially the brook trout. Introduction of the same species into new waters or reintroduction into former habitats is also being considered. But the Vibert box method is truly made to order for application by small groups of dedicated fishermen who want to see their local waters improved. The method is simple enough and so inexpensive that it can be used by average fishermen if they are willing to take the time to develop a local program. It is also extremely interesting and enjoyable group work.

If many groups across the country undertake this program, it could have an incredible effect on the quality of our fishing for trout and salmon. Not only would it provide better stocks of fish and higher angling quality, but it would certainly infect those that work with it as it did me: it would give each person a deep insight into the true meaning and value of our fish stocks, and it would reverse our downward spiral of less sport and more cost, to more sport and less cost.

The Federation of Fly Fishermen has adopted the Vibert box system as one of its highest priority programs. The organization is financing research and giving unlimited assistance to all interested groups that wish to start a local Vibert box project. Presently available on request to FFF is my detailed booklet explaining all aspects of the Vibert box stocking system, and where you may purchase them and obtain eggs. The booklet goes into far more detail than I could present here, or that has ever been reported elsewhere. Also, for organized fishing clubs, similar interest groups, or state game departments, there are two tape-narrated 35mm color-slide shows available on special request from FFF. The first show, "The Vibert Box Stocking System," is a twenty-minute special about the method; the second, "How to Use the Vibert Box," is a sixty-minute detailed instructional show that gives invaluable information for successful application of the Vibert box. (See address at back of this book.)

When Dr. Vibert conceived his hatching box, he gave us a means of undoing some of the wrongs we have allowed our trout and salmon to suffer. The box allows nature to have about eight times more to work with in her production of a stream full of wild trout. Those are good odds. They warrant our assistance.

Successful natural reproduction or artificial stocking methods are the responsibility of everyone who loves *live* waters. By learning as much as we can about the life of a stream, we can develop a keen understanding and deep appreciation for the treasure of fish that exist there. And then, when the day arrives that we all consider a game fish far too valuable to catch only once and kill for unnecessary food, then the fish of our waters will be able to populate and grow to normal limits. Sportfishing methods, workable regulations, sensible harvest, strictly obeyed laws, and deep respect for the privilege of angling will all drastically reduce the declining state of our national fishery. Evidence not only of a changing interest but also of a reverse in the tide is already apparent—but there is still much work to be done.

A recent study by Montana Fishery biologists has revealed some extremely important facts and insights in regard to stocking hatchery-reared trout within a healthy wild trout fishery. Dick Vincent's article "The Catchable Trout" reveals some significant findings. This report covers the impact and practicability of dumping tame fish in wild trout streams to "boost" tourist-angler success. The study accurately documents a two-fold depression on the stream.

First, these hatchery fish make only a brief contribution to angling success and perish before they can make any significant contribution in growth or reproduction. The river environment soon adsorbs what are not jerked out by meat fishers following hatchery trucks. Second, these thousands of tame fish overcrowd the territories of wild stocks holding in the same water. This drastically upsets the territorial needs of wild fish and they abandon the area in search of better areas. Then the hatchery fish dwindle away from various mortalities, and the addition of more stocks repeats the dilemma. But the wild fish do not return in original number, and the river is thus diluted of its only real fish. Finally, reproduction is lost, and angling quality falls off drastically.

This study painfully reveals what many of the serious anglers and fishery managers have long suspected: stocking tame fish actually *depletes* a good water, fishing pressure or not. Streams allowed to maintain themselves usually will do just that with sensible management. Similar areas, the Madison River and Odell Creek, reflected dramatic population increase when allowed to support themselves with natural reproduction.

Where natural reproduction from native stocks is possible, it and the Vibert box method are the only practical and meaningful methods of maintaining healthy, quality fisheries. Only artificial or marginal waters justify the "put and take" method. But what of good waters exposed to high angling pressures? *Good management* on any quality water can stretch a lot of angling hours from each fish. Fish for sport and fun and no-kill ideals have proved themselves in such waters as Armstrong, Yellowstone, Big Spring, Hot Creek, and others across the country.

The joys of taking treasure from a stream—and of returning to it that same treasure for the future. *Dave Whitlock*

WARMWATER FISH

The various warmwater species—particularly the sunfish family, which includes largemouth bass, smallmouth bass, spotted bass, and such panfish as bluegill, redear, green sunfish, rock bass and crappie—are extremely popular and important sport fish throughout North America. Good management of these sporty species is quite different from that used with the coldwater salmonids. Natural spawning in healthy environments usually is adequate to assure a plentiful supply of fish to populate most streams and impoundments.

Spawning occurs in the spring or early summer, dependent upon water temperature and stability; usually the ideal water temperatures range from 65° to 75° F. Most sunfish are nest builders. The bass are solitary nesters, while most panfish "bed" in small to large groups, sharing a general ideal area. They choose a sheltered area outside the main stream flow; there the male will construct a circular fine-gravel-and-sand nest by cleaning and digging the bottom with his tail and body. He then invites the first ripe female to join him either by special antics or nipping at her fins until she moves into the nest area. There she will lay her eggs, usually tens of thousands, depending upon species, age, and size. Sunfish eggs are so tiny they are almost invisible; many thousands more are carried by a female sunfish than by a trout of similar body weight.

As the female lays her eggs, the male fertilizes them with a cloud of milt. Soon after the eggs are deposited, the female leaves the nest and returns to feeding and normal stream life; but the male will guard the nest from intruding eggs thieves—especially minnows, suckers, immature panfish, and even some insects. His movements also help assure that adequate oxygen gets to the live eggs. Despite his best efforts, though, many eggs are eaten or destroyed.

As the eggs hatch, usually taking several weeks under ideal temperature conditions, the tiny fry emerge and form a moving black cloud or school. At this time, the fry are constantly attacked by larger fish and aquatic insects. Even the father will eat many fry, breaking his many weeks of fast by enjoying a meal of his own brood. But the odds are with the numbers, and thousands of fry manage to escape to begin their normal stream life. Usually the bass mature at two to four years of age, panfish

Bluegill spawning on nest. Most panfish "bed" each spring in such community spawning areas, sometimes beds will include fifty to one hundred nests. Such togetherness provides sharing of limited ideal areas, protection, but also encourages hybridization between similar species, such as bluegill and redear.

a year earlier. Maturity also depends much upon the latitude.

Natural reproduction is undoubtedly the most preferred and successful method for warmwater species. Only under a few circumstances do most states stock warmwater streams, rivers, or lakes these days. Sometimes natural disasters such as drought, or unnatural ones such as man's pollution, kill out native stock. Then restocking is called for. Usually this is only needed over a one-to-three-year period to regenerate natural reproduction.

Since the sunfish family reproduces by natural spawning only and demands live food from birth, hatchery methods differ greatly from those employed by hatcheries for trout or salmon. Usually adults of both sexes are placed in natural holding ponds or raceways and allowed to mate and spawn on their own. After spawning occurs, the adults are removed and the young fry are allowed a predator-free sanctuary to begin their development. Such brooder areas are provided with natural live foods. Bass are especially reluctant to accept prepared foods, and little bass or panfish feed well only when allowed to forage on living forms such as aquatic insects, crustacea, and minnows. Pellet feeding has been successful for certain species of panfish and hybrids. After sufficient growth has occurred, usually to fingerling size for bass (2½″ to 4″), smaller for various panfish, the brooder areas are drained and young stock removed for stocking deficient waters.

This may seem simple and economic when compared to the production of cold-water fish. But this is not entirely true. Many warmwater sun fish, especially the smallmouth bass, spotted bass, and rock bass, are reluctant to reproduce for the whims of man. Under most circumstances only a small percentage of pairs nest successfully except in the most elaborate hatchery operations. Also, providing ample live, microscopic food for the developing fry can be an expensive nightmare. It is also quite common for the young to turn cannibalistic and literally eat themselves out of existence.

Spawning largemouth bass over nest. Male is always smaller than female in bass sexes of similar ages.

Stocking bass seldom provides much increase in take-home poundage where natural populations exist in balance with the water system. It is far better management to improve the environment by removing such competing species as suckers, shad, carp, and panfish. Concentrated fishing pressure on desirable adult bass removes natural predator balance controls on these lesser species. This allows the lesser species to overpopulate, thus removing important cover and food for the young bass. It is a vicious circle that fishery people must contend with and one not easily or quickly solved. Usually they resort to total removal of all species, and then rebuild first cover and then food forms by fertilizing and replanting vegetation. This sets the food chain into motion by providing enrichment of basic simple food forms.

Now the stage is set properly for ideal growth development and the water is ratio stocked with several compatible species of fish. Usually largemouth bass, hybrid panfish, bluegill, crappie, and channel catfish are planted. Whatever is stocked, a period of fast growth and excellent spawning success follows for several years. This continues until imbalance due to improper fishing harvest or accidental introduction of undesirables occurs.

As with cold-water species, warmwater game fish populations respond to good conservation practices. The main problem is to keep the balance between the food sources and predator species. If adult game fish are overharvested by the fishermen, the balance soon tips off, especially if trash fish are ignored by such anglers. The removal of prime fish eliminates their spawning potential and assures trash fish and prolific panfish freedom to go at the food and space.

Some state authorities restrict kills by setting minimum size limits, maximum daily bag limits, and a closed season during the spawning months. Though this is effective management, it seems to be extremely unpopular these days. The popular theory now is that such practices are not practical, and that reproduction and resident populations are usually adequate or far above adequate. This may be fairly good thinking for large southern impoundments or underfished small lakes, but it is seldom if ever true for the streams. Lakes provide considerable sanctuary for fish populations and far less severe environmental extremes than streams. If we cannot convince the custodians of these waters to have separate restrictions and management policies for running and impounded waters, then it is up to each of us to practice good conservation on the streams.

Fish each of these unique flows with consideration, and release most if not all prime game fish. Don't introduce trash fish by using live minnows as bait. Use of artificials, especially one-hook lures, prevents damage to gills or stomachs, which result in death even if the fish are released. If there seem to be a lot of small panfish or rough fish present, remove as many as you can.

While most warmer-water fish seem to be doing better than the trout in our streams, I am deeply concerned about the fate of one wonderful fish—the smallmouth bass. It seems to thrive in those lakes where it occurs naturally or has been successfully introduced, but in running waters, especially throughout the East, Midwest, and Southwest, it is threatened. Stream smallmouths are being exposed to dangers—dam building, industrial pollution, and overharvest—that threaten their eventual extinction. The smallmouth is literally disappearing from many waters today! The uncompromising nature of these great game fish put them into even more jeopardy than most trout. They reproduce only in nature, seldom being successfully raised in hatcheries. Their growth rate is extremely slow, maturity seldom coming before the fourth year: a typical Ozark stream smallmouth is only about ten to twelve inches at that age. Of course in lakes or very large rivers the growth rate is better.

Many of the smallmouth rivers are being dammed, forming large hydroelectric flood-control lakes. The majority of these lakes prove too warm or muddy to fit the smallmouths' needs, and these stream fish do not make adequate adjustment to lake life and soon disappear. It is a dilemma that seems almost unsolvable when one considers the increasing fishing pressure, the exploiting uses of these flowing waters, and the dam builders' future plans.

It is my hope that the authorities can be encouraged to recognize these problems and take the necessary steps to save our most thrilling and unique warmwater game fish—the stream smallmouth bass. I rate him at the top, well above the brown or rainbow trout for sport. The smallmouth is the perfect indicator of a sick or healthy river. If we lose this unique species, we might well lose much more in the process.

7

Group Action and Available Support

R. P. VAN GYTENBEEK

The answer must come from individuals like you and me, for civilization with his governments and establishments is shaped by the forces of individual desire.
CHARLES A. LINDBERGH

I am often asked, "What can *I* do?"

Sometimes the question is posed by an individual wishing to volunteer; more often the inquiry is made in frustration, and might better be phrased, "What can *one* man do in the face of the onslaught on our streams and their fishery resource?"

There is a great deal that an individual can do for his favorite water: litter removal, a few rocks judiciously placed, planting on a recently eroded section of stream bank, or calling the proper officials upon observation of a new source of pollution or an act of physical desecration. These actions, undertaken by each fisherman and coupled with similar action by other individuals, can make a significant contribution.

Yet the real answer to the question is: *"Get involved with an organized group that is doing something."* For within this context, the individual angler's effort will gain immeasurably more strength, and his "individual desire" will become a shaping force for the preservation, restoration, and enhancement of our stream fishing.

Both *physical* and *conceptual* work of great importance has been accomplished by such groups.

Destruction of the environment threatens the future of angling. It is a common goal of the Federation of Fly Fishermen and Trout Unlimited to resist that destruction and repair damage already done. The Executive Boards of both organizations accordingly have adopted the following resolution, which now becomes their official policy:

A Resolution . . .

As two organizations acting in concert to safeguard our sport fishing and our natural resources, we of the Federation of Fly Fishermen and Trout Unlimited, do pledge our appreciation and support for each other's programs in the philosophies of ecology, environment and conservation; we further pledge to formulate plans that will advance our fishing sport as well as preserve and enhance our environment; to pull together in an orderly and effective way the talents and dedication of individuals in each organization; to further let it be known that as members, chapters and clubs, we will work together to actively protect local water resources and to improve water management practices in all areas. Finally, we lend our support and efforts to all other organizations who are striving for the same environmental qualities. Any assault on the quality of our environment is often an assault on all.

James R. Eriser, President
Federation of Fly Fishermen

Otto H. Teller, President
Trout Unlimited

This resolution (and the photo montage), first published in the *Flyfisher,* the official publication of the Federation of Fly Fishermen, pledged mutual cooperation by the Federation and Trout Unlimited in the fight to preserve the natural environment on which the sport of angling depends.

PHYSICAL CONSTRUCTION PROJECTS

Physical improvement of our waterways is most important, and much can be accomplished in this area through detailed planning, dedicated leadership and, in many cases, cold hard dollars.

Let's assume that you and your fishing buddies have selected a section of stream that you feel should be improved. You see that the bank is broken down, that trash is in the water, and that some sections virtually dry up in the heat of high summer. You *know* that it can be improved—or do you? At this point, you are operating on barroom theory. Your first step must be to get the facts. Is your proposed approach the best one? Is the stream worth saving? Can it be saved? Can it be improved? With all the effort that you propose to put forth, give yourself a reasonable chance of producing a significant improvement.

As you contemplate these various questions, you may feel overwhelmed. Don't! Many agencies at all levels of government can give you the necessary facts, and they probably stand willing and able to assist you. But before you contact them, take one more important preliminary step. Find out who owns the particular stream section you wish to improve. If you're in the West, check on the ownership and use of the water rights. If the river is in private ownership or the water rights belong to someone, check with these individuals, outline your plans, and gain their permission to work on the stream. If the water is under the ownership or management of a public agency, then get their permission. Once you've gained this, pick out the agency that seems most likely to be able to assist you. In most cases, this will be your state fish and game department. While some states maintain a stream-improvement section within their division of fisheries, most presently do not, and it will probably be necessary to find the regional fisheries biologist or the man or section directly responsible and discuss with this individual or group your plans, hopes, and desires.

If he is well informed, he will probably go to his files and dig out the latest base-line data on the stream in question. If you live in a state where this data is not yet available, it may be necessary for the state fish and game or other agency to go out and survey the stream, establish base-line, physical, and chemical data (history, population profiles, and the like), and then

Missouri fishermen demonstrate love of their sport by collecting trash from Dry Creek.

assist you in drawing a plan to accomplish the proposed stream improvement. All of this will undoubtedly take time. It may be tedious. Some of your group will become frustrated during the fact-finding process. But if you're going to do it, do it right!

Finally, you have the necessary data in hand. You have a plan on paper. You've identified those improvements that can most benefit the stream and you're ready to go to work. In the next, or organizational, stage, there are probably three critical components: manpower, material, and equipment. In this day of environmental awareness, manpower is often the least of your problems.

Do not try to implement all of your plan with your own small group. *Involve others.* No matter how enthusiastic your particular group may be, on that cold, gray, spring day on the stream when the project starts, you'll be lucky to get 15 to 20 percent of them actually out on the project. It's amazing how many screens have to be washed, garages swept, golf dates kept on such days! Enthusiasm in a cozy meeting room has to be transferred to the cold, unfriendly waters of a trout stream during inclement weather—the kind of day on which, with uncanny regularity, improvement projects usually are scheduled. Probably the best source of enthusiastic additional manpower is the youth of our country. The Boy Scouts, Future Farmers, the Brotherhood of the Jungle Cock, ecology classes in the local school district and nearby college campuses are sources to tap. Young people are hard workers, but you should try to put a little fun in the project for them and, in the case of most groups, some education too. Don't plan to work them all day long. A couple of hours of hard work separated by instruction and some good refreshments will produce a second period of worthwhile effort later in the afternoon and guarantee that the same youngsters will be back to help another day. When you're instructing them during the break periods, be sure to emphasize what it is that they're doing and what it will accomplish.

This type of work party should be carefully organized so that people are not standing around wondering what to do. You will have spent a tremendous amount of time and effort getting the

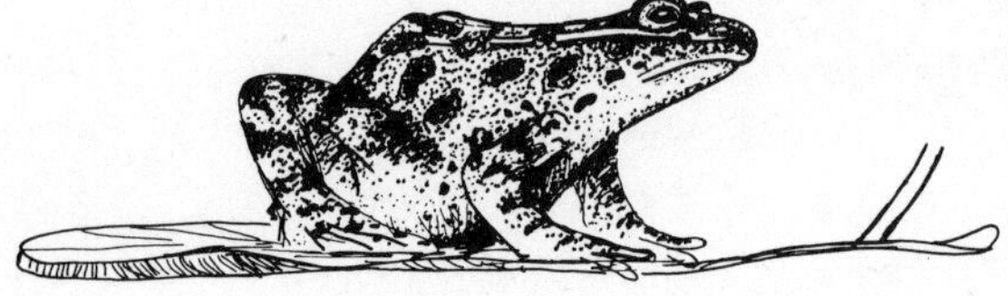

people and the stream together; see that everyone knows what to do, and that they have the tools to do it with. One of the problems that invariably occurs is the difficulty in getting the needed materials to the site. Do this beforehand, and do it smoothly. There is nothing worse than getting a group of eager volunteers to the site only to find that because they lack some material or equipment, they are unable to work properly. Assuming that your efforts will take place on weekends and holidays, there are invariably civic-minded construction firms willing to donate equipment and often operators to assist you. If operators are a problem, a short session with a local labor union may uncover a number of individuals who are willing to operate the equipment as a donation to a worthy project.

Good media coverage will maintain enthusiasm for your project. On the first day, ensure that radio, television, and newspaper people are informed; nothing makes better press for a Sunday supplement than a picture of youth working on a conservation project. It's surefire material for the paper and also for TV newscasts. Good coverage helps morale and advertises your program. Probably Monday morning will bring calls and inquiries from people and groups who wish to get in on the action.

Does this pattern work? Three recent projects carried out by TU chapters prove that it does. The Grays Run project in Pennsylvania is an example of conservation in action. In the summer of 1966, TU's Susquehanna Chapter, located in Williamsport, Pennsylvania, decided to try to undo the ravages of floods and poor land use that had begun to destroy a lovely native brook-trout-producing stream in their area. Careful preplanning with the state fisheries personnel, Department of Lands and Forests, and private fisheries experts such as Dr. Alvin R. Grove, National TU Director and nearby resident, helped gain approval of their plans from all the necessary agencies. The chapter next involved the West Branch Council of the Boy Scouts. By the summer of 1967 a massive stream-improvement program, which was finally to involve 45,000 hours of labor, had begun. The project was essentially complete in the early fall of 1969, and later that year the TU chapter and the West Branch Council were recognized for their outstanding effort when they were presented with the United States Department of Agriculture's Gold Seal Award. The award recognized

Grays Run as the top citizen conservation program undertaken in the nation during 1969.

This project is an excellent example of how it should be done. Though there were many who wished to proceed and became indignant that work did not start sooner, a year of preplanning went into the effort before the first stone was turned. Fortunately, the chapter leadership was committed to a well-conceived and approved plan. The Boy Scouts, who contributed the major portion of the labor, worked half days; the other half day was devoted to learning how to fish, acquiring greater appreciation for the natural lands surrounding the stream, and having fun under the supervision of their leaders and the members of the Susquehanna Chapter.

After a couple of years of revegetation, it is very difficult for the casual angler to observe the work that has been done; and surveys by the Pennsylvania Fish Commission indicate almost a doubling of the native trout population in Grays Run.

In addition to the fact that the fishery was vastly improved, the chief of the U.S.F.S. may have made an even more significant point during the dedication of the project when he said: "In the years to come, it will be very difficult for highway builders and other despoilers to damage this particular valley and its stream because there will be twenty thousand citizens in the area who each have a little bit of blood, sweat and tears in this stream and who have a real stake in its future."

In Denver, Colorado, the Cutthroat Chapter of TU found that the lower eight to ten miles of a small tributary to the Platte River, Bear Creek, had been channeled and visited by the normal collection of junkyards, poorly conceived residential areas, small feedlots, and much else. Again, a small group of men believed that the stream could be restored, and many were ready to charge off immediately to throw in stream-improvement structures, stock the area with trout, and sit back to admire their handiwork.

Fortunately, clear heads prevailed and over a year of planning went into the program before the first gabion was placed, the first willow planted. The chapter worked closely with the Colorado Game, Fish and Parks Department. They tested the water flow, quality, temperature, and oxygen content throughout an entire year. A speakers' bureau was formed and local schoolchildren and their instructors, Boy Scouts, and others

Stream-improvement projects, such as this one on Bear Creek by Trout Unlimited members, can decisively improve the quality of our rivers. Essential ingredients to the success of such programs are careful planning, hard work, and cooperation. *Photo courtesy of Phil Key*

were told about the program, sold on the concept, and brought into the planning of the project. Those members of the chapter who had engineering expertise carefully mapped the stream and together with a state fisheries biologist conceived a program to rebuild the stream. The difficult logistics of bringing young people, equipment, and material together at the same place and the same time was accomplished. After two years of effort a number of miles of stream have already been rebuilt with stream-improvement devices and bank plantings; trout again swim in Bear Creek and plans have been mapped to improve and restore the remaining damaged sections. Good preplanning, good organization, and a strong-selling job have brought back a stream that was fishless for some thirty years.

Not far north of Denver, between Fort Collins and Greeley, Colorado, a large section of the Poudre River, which has been fishless for decades, is also being rehabilitated. Here, again, over two years of effort have gone into the planning for a major stream renovation project. The Poudre's major problems were water quality and quantity. Because of irrigation demands, the

Members of the local chapter of Trout Unlimited shock the Poudre River, as part of a well-planned restoration project. *Photo courtesy of James Redder*

stream flow had been greatly diminished and that water which was used for irrigation came back to the stream carrying heavy silt loads as well as chemicals washed from the surrounding farmland. The city of Fort Collins was found to be providing inadequate sewage treatment and a number of other polluters were also identified. Two chapters of TU, working in conjunction with the United States Geological Survey, the Environmental Protection Agency, and others have been able to set up a timetable for cleansing the stream as it flows through Fort Collins, thus providing an essential ingredient needed for the thirty-odd-mile piece of water between the two cities.

During the planning stage for the rehabilitation of the Poudre River, a very fortuitous happening occurred. The Kodak Corporation announced construction of a new plant on the banks of the Poudre in the little town of Windsor, approximately halfway between Fort Collins and Greeley. At first this

A team of Trout Unlimited members shocking Colorado's Poudre River. *Photo courtesy of Steve Powus*

announcement seemed to the chapters as though it would most certainly be the death knell for the Poudre, but direct contact and discussions with executives of the Kodak Corporation brought about an agreement that in the end will probably prove to be the single most important factor in reestablishing the trout fishery. The company, which has shown its environmental concern in many ways, had decided to put in a very high-quality treatment facility to process the water it used. Here was the source of pure, clean, cool water that the river so vitally needed, especially in its lowest sections. The corporation gave the chapter permission to work on its land. After two years of planning, discussion, and work with the necessary agencies and with the Kodak Corporation, the two TU chapters stocked wild brown trout fingerlings obtained through the courtesy of Wyoming Fish and Game in the Kodak section of the river, a section which a survey proved to be an acceptable location.

On the West Coast, we find another excellent example of cooperation among industry, state agencies, and a conservation organization. It's the well-publicized Hat Creek Project, brainchild of the California Council of Trout Unlimited. The project started with a pristine but rough-fish-laden stream and the premise that the people of California wanted quality angling for native fish, not just put-and-take for small hatchery-raised trout.

Hat Creek, in Shasta County at the northern end of the state, had once been noted for its brown and rainbow trout fishery. The stream meanders 3.2 miles through lush meadows from Pacific Gas and Electric Company's Hat 2 Powerhouse to Lake Britton, created by a power dam about four miles downstream.

Over the years, Lake Britton has become infested with large numbers of rough fish, such as suckers, hardheads, lampreys, squawfish, tule perch, and green sunfish. It is a warm water nursery for nonsport fish. The rough fish moved into Hat Creek and comprised about 93 percent of the population of the stream to the virtual exclusion of native trout. Trout fishing became impossible and the trout's struggle for survival was a losing proposition.

The competition for food and habitat between trout and rough fish was tremendous. Run-of-the-mill hatchery trout reared for the put-and-take programs could not survive in Hat Creek because of the presence of a protozoan parasite, *Ceratomyxa shasta,* which proves fatal to nonnative or specially bred disease-resistant trout, such as the Mount Whitney strain of rainbow trout.

By weight, the suckers, hardheads, squawfish, and such comprised 97 percent of the fish life in this stretch of Hat Creek. While some of the native trout reached one to two pounds or more, their fry didn't stand much chance for survival in such a competitive atmosphere.

Hat Creek was becoming a wasted stream. Such an ideal, classic mid-elevation stream should not be allowed to deteriorate, California Council of Trout Unlimited members thought. Jim Adams, a PG&E aquatic biologist, Andre Puyans and the late Joseph Paul, all TU members, thought the problem could be remedied by chemical treatment of the stream to rid it of the undesirable fish population. Erection of a barrier to prevent reinfestation from Lake Britton would keep the trout neighborhood exclusive. Adams, working with California Department of Fish and Game biologists, recognized that the *Ceratomyxa shasta* problem couldn't be licked so the stream would have to be replanted with native and disease-resistant trout.

Joe Paul was a dynamic and forceful public relations entrepreneur and spearheaded a joint project to restore Hat Creek. He approached PG&E, the state's largest energy utility, which controls Hat Creek for hydroelectric generation and owns the lands adjacent to the stream. In response to Paul's pitch to restore Hat Creek, PG&E management pledged cooperation and assumed responsibility for designing and supervising construction of the barrier. The company also agreed to contribute $2,-500 toward its cost and to grant an easement to the State of California for the barrier and fishing access.

The California Wildlife Conservation Board provided $13,-000 for barrier construction and agreed to erect signs on the project. (WCB funds come from parimutuel betting.)

The Department of Fish and Game agreed to develop information to permit evaluation of the project, to chemically treat the stream following construction of the barrier, to stock it with native and selected strains of trout and to manage the stream as a wild trout fishery. They proposed a reduced bag limit of two fish and no tackle restrictions to the California Fish and Game Commission.

California members of Trout Unlimited agreed to raise $10,-000 (subsequently increased to $12,000) for a trust to fund scholarships for graduate fishery biology students working on a six-year research program in connection with the Hat Creek Wild Trout Project.

When the final agreement between the California Department of Fish and Game, the California Wildlife Conservation Board, Pacific Gas and Electric Company, and the California Council of Trout Unlimited was signed in 1967, the product was hailed as a union of governmental agencies, a public utility, and a conservation organization to work together in restoration of a resource of significant public value.

Before construction of the barrier began, the stream population was sampled by the Department of Fish and Game with help from TU volunteers, PG&E aquatic biologists, and graduate fishery biology students of Humboldt State College at Eureka. They used electro-shocking devices to stun but not otherwise harm the fish. Trout were marked and released for later capture. The resulting population estimate proved feasibility of the project.

Before chemical treatment, the electroshocking was done again, this time to recover as many native trout as possible for brood stock. They were taken to nearby Crystal Lake Hatchery and were kept separate from other fish.

The one day operation in October 1968 to chemically treat 3.2 miles of Hat Creek resulted in the destruction of 12,985 pounds of rough fish, mostly suckers, according to figures compiled by the Cooperative Fishery Unit at Humboldt State College.

Four days later the trout removed in the electroshocking operation were returned to the stream, which was then closed until the late May general trout season opening in California.

Department of Fish and Game figures show that by the end of 1972 two-thirds of the fish caught in Hat Creek were being released by anglers to be taken again. Some fish have been caught as many as six times.

And a report issued by the California Department of Fish and Game shows that the average rainbow caught during 1970 weighed .7 pounds, up from .5 pounds during 1969. Brown trout weighed an average of one pound, up from a 1969 weight of .6 pounds. The growth rate is directly attributable to improved habitat and reduction of competition for food.

G. Ray Arnett, director of the California Department of Fish and Game, says, "A secondary payoff from the Hat Creek Project is the wealth of information about the biology of the fish and the composition of the stream." He calls attention to the relationship between trout and rough fish, which will have significant implications for the future management of trout fisheries.

Virtually every inch of Hat Creek in the wild trout management area has been studied by graduate fishery biology students. They have produced a valuable and extensive scholarly literature on the stream and its inhabitants, a basic wildlife research effort which makes the project that much more significant.

California now has twenty-two streams marked for management as wild trout fisheries. The lessons learned at Hat Creek may well be applied in many cases to successful management of the streams included in California's plan, and in other areas.

One federal program, which has been most important to conservationists for the last thirty-odd years, the Rural Environmental Assistance Program (REAP) has been temporarily defunded by the Nixon administration. However, it is anticipated that Congress will reestablish it in 1973, and if not then, certainly in early 1974. This important program provides cost sharing by the federal government with the landowner for approved conservation programs in rural areas.

Preparing to release a huge trout in Hat Creek. Two-thirds of all trout caught are released. *Pacific Gas and Electric Company*

A fish barrier on Hat Creek keeps the trout neighborhood "exclusive." *Pacific Gas and Electric Company*

Al Long with two trophy browns from Hat Creek. Proper management and cooperation can create this kind of fishing in *your* area, too! *Pacific Gas and Electric Company*

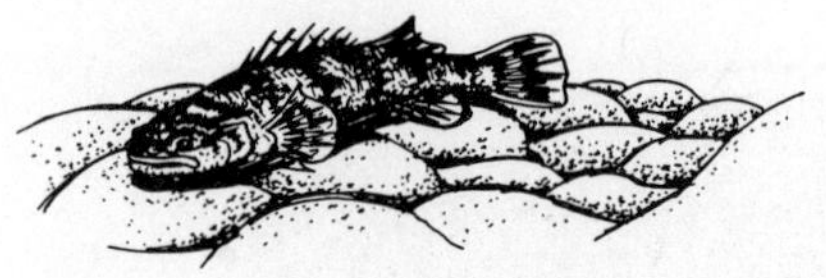

You may ask, "What's that got to do with us?" A good question.

Let's look at what it had to do with the Western New York Chapter of TU on the Wiscoy River in western New York. This fine trout stream was suffering from many ills, paramount of which was the degradation of its banks from overgrazing and highway construction—which in turn created silt and thermal pollution. The Western New York Chapter surveyed some thirty miles of this stream and were able to pinpoint those areas that were creating most of the problem. They then contacted the landowners who owned these sections and explained their desire to stabilize the banks, fence out stock from the majority of the bank, and reestablish streamside planting. The landowners agreed with the plans and made application to the Soil Conservation Service for the REAP funds. The federal government provided 80 percent of the necessary funds while the other 20 percent was credited to the chapter for its labor. In other words, each man-hour of effort was credited for a certain dollar figure and the labor was to become (as it always is) a measured percentage of the overall cost of the project.

Thus, without any dollars out of pocket, the chapter was able to use the federal matching money for materials, put up their own labor, and accomplish the project. Miles of the Wiscoy banks have now been planted, stock has been fenced out, banks stabilized; the stream, under study by the New York State fisheries personnel, has already proved to have been greatly enhanced.

Of all the various federal programs, the REAP program probably represents the greatest potential for the concerned stream conservationist. Those who are interested in details, should get in touch with their nearest Soil Conservation Service office.

CONCEPTUAL PROJECTS

Other projects have involved little if any actual physical effort. Instead, they require a great deal of *selling*—that is, mental and verbal effort rather than physical effort.

Trophy fishing, fish for fun, hook and release, fly-fishing only, artificials only are concepts near and dear to many serious trout fishermen. Organizations such as the Federation of Fly Fishermen, Trout Unlimited, Cal Trout, and others have attempted to promote this type of regulation with varying degrees of success. Most of us who are working with the problem on a daily basis know that no more than 20 percent of the anglers in any given area, usually a far smaller percentage, are truly interested in imposing these so-called quality regulations on themselves. A call for fly-fishing only, for instance, immediately brings down the anger of the worm dunker and lure slinger who feel that they are being unfairly legislated against; and in truth a study of the available research indicates that at least the lure fisherman is justified. Specifically, fish caught on single-hook spinning lures show at least as great a survival under a hook-and-release system as do fish caught on a fly. The bait fisherman does not fare so well in this argument; the same studies show that fish hooked and released on bait invariably show a fatality rate higher than 50 percent. This is opposed to mortality from single-hook lures and flies of 5 to 8 percent.

It makes sense to understand a sound basic management philosophy such as TU's North American Salmonid Policy,

This booth at a sportsmen's show was a joint project of several different fishermen's and conservation groups who pooled their time, money, and membership to a common goal. While none of them was able to sustain this kind of effort by themselves, together they were able to secure support, memberships, and donations. *Scenic Hudson Preservation Conference*

which requests that each management agency fully inventory those waters that are under its control, categorize the various waters, and evolve and implement a management program based on those categories. In this way streams capable of producing a population of native trout will be managed so that only the harvestable surplus will be taken without the high cost and the detrimental side effects of stocking. Streams that are capable of growing and holding over fish will be stocked only with fingerlings, and finally those waters capable of carrying fish for a short period of time—but incapable of either carrying them through the winter or reproducing them—will receive infusions of hatchery-reared catchables. Once these determinations are made, it is possible to concentrate intelligent applicable management philosophies on one category or another.

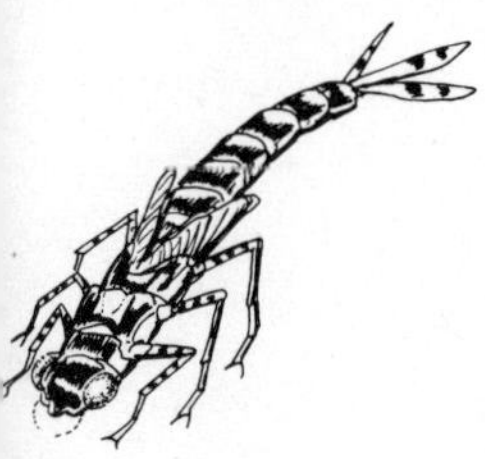

On the wild trout waters, where you want to take no more than a predetermined harvestable surplus, limited kill methods *must* be applied to assure that the kill does not exceed the surplus. Here the interested individual or group should think diplomatically. For instance, where fly-fishing only may be onerous, who would disagree with a trophy-fishing regulation or fish for fun? After all, everybody likes the fun and everybody would like a trophy, so if that's the terminology you apply you may be able to do so and still have the same restrictive methods, but under a different guise—one acceptable to the public, and thus to the fish and game commission.

Many of us from the northeastern part of the country well remember the years and years during which the Theodore Gordon Flyfishers and others in New Jersey and New York tried, to no avail, to place fly-fishing-only regulations on a substantial stretch of the Beaverkill River. Every year the request met with charges of discrimination. It was not until someone had the bright idea of calling it a trophy-fishing area that the regulations were adopted. And while other methods are allowed, the no-kill and large-size requirements have discouraged bait and most spin fishermen from using the area. Thus, through the simple expedient of applying a more salable label, we really have what all of us were trying to get for years and years without success. And the no-kill stretch provides exceptional fishing today. On Waters Creek in northern Georgia, the Chattahoochee Chapter of TU has been interested in supplemental feeding of naturally produced wild trout. Numerous private clubs in the southeastern part of the country use this method to provide truly superior fishing on their privately controlled waters. The North Carolina Council of TU has induced their department of natural resources to try this method on public waters, and Georgia felt that they could also. The head of Georgia's fisheries division, Leon Kirkland, was invited to go on a weekend "show me" trip to North Carolina, where he was shown a private stream and the quality fishing it provided under a supplemental feeding program. As the result of this trip and other spadework, Georgia Fish and Game became most enthusiastic and designated a small stream, Waters Creek, as a trophy-fishing area. The chapter, for their part, agreed to assist the fish and game department in patrolling, signs, and cleanup of the area—and this certainly helped the department to reach their decision to go ahead. During its first year in operation the project has proved most successful and popular with Georgia fishermen. It is anticipated that additional projects will be implemented.

Armstrong Spring Creek near Livingston, Montana, is probably the best-known Spring Creek fishery in the country. It remains in the public domain today as the result of action by a small group of Livingston area anglers. A few years ago, after the tradition of leaving the water open to the public for fishing by permission, Allan O'Hare, the ranch owner, told his friend Dan Bailey that a group of private individuals had made him an offer of $12,000 per annum to lease Spring Creek. O'Hare indicated he would have a difficult time turning down this lucrative offer. Dan asked for time, talked to the national headquarters of TU and the local Yellowstone River Chapter of the organization. The decision was made to lease the river and try to raise enough funds on a yearly basis to keep it open to the public, under the sponsorship and management of the Livingston Chapter of TU. Rancher O'Hare agreed to a yearly lease at $6,000, or half of what he had been offered by the private group. Through the efforts of Dan Bailey, the Yellowstone River Chapter, and TU National, money has been raised each year from interested individuals, tackle manufacturers, and others to service the lease. The chapter has worked on stream improvement, access points, and parking areas—and today one of America's greatest trout streams is still open to the public because of the dedicated efforts of a few individuals.

The sign at Armstrong Spring Creek. *Photo courtesy of Dan Bailey's Fly Shop*

The Big Spring on Armstrong Creek. *Photo Courtesy of Dan Bailey's Fly Shop*

Some of the superb smaller water above the Big Spring. *Photo courtesy of Dan Bailey's Fly Shop*

Some conceptual projects may be large scale. Witness the Colorado TU Council's efforts to preserve and restore the Platte River as it flows through Denver. A number of years ago this chapter decided that the Platte could once again support trout in Denver, and studies of the river proved this to be true. A two-year battle with the commissioners and legislative bodies got the river's classification upgraded and the polluting industries and communities put on cease and desist orders. The chapter was in the forefront of those supporting construction of the Chatfield Dam, a Corps of Army Engineers flood-control project that guaranteed a bottom-drawn minimum flow of clean, clear, oxygenated water. Having achieved all this, the interested Denver fishermen were about to sit back and enjoy the fruits of their labor when they found out that not only was the Corps planning to build a flood-control project for over $200 million of the taxpayers' money, but now they also wanted to get their hands a little deeper in pork and channelize the river for miles downstream of the dam, destroying all hopes for the reconstituted metropolitan fishery.

Quickly mobilized again, TU worked with their congressman, federal agencies, the Bureau of Outdoor Recreation and the community of Littleton through which the stream flowed. The congressman was able to have the Corps funds diverted from the purpose of channelization to flood plain acquisition; the Bureau of Outdoor Recreation provided matching funds; and the chapter, through door-to-door solicitation, and the help of others, was able to convince the citizens of Littleton to pass a bond issue which, in turn, provided the matching funds for the B.O.R. monies. Years of effort—involving work and coordination with local, state, and federal agencies, a political campaign, a public relations campaign, and many other efforts—will finally result in a fishery where none has existed for thirty years.

There are hundreds of additional examples of what can be accomplished both conceptually and physically. Most major undertakings, like one from the northwestern part of our country, involve both.

The barren rivers program in the state of Washington was the brain child of TU's Western Regional Director, Chuck Voss. During the last few years the coho salmon hatchery program of the state of Washington has been fabulously successful. So successful, in fact, that hundreds of thousands of salmon were being wasted because the department neither had the hatchery facilities nor the transportation facilities to hold these fry or to plant them in a suitable river. At the same time, Voss discovered that there were many rivers suitable for salmon that did not have either a natural run of salmon or steelhead or a resident trout population. He reasoned that these barren rivers, if stocked, could open up a whole new fishery opportunity for the hordes of northwestern anglers. He quietly promoted his plan with various state agencies and finally gained the approval of the state of Washington for his barren rivers program.

As Chuck conceived it, this would involve assistance in spawning and stocking the surplus fish into barren rivers under Fish and Game supervision. Many of the Northwest Steelheaders group, which Chuck serves, have float boats, trailers, and pickup trucks, and a method was worked out whereby plastic garbage cans could be aerated for a sufficient period of time to allow them to be towed on trailers or taken in the back of pickup trucks to the barren rivers. On weekend after weekend, convoys of cars with trailers and pickup trucks, all toting plastic garbage cans with portable aerators, could be seen moving from Washington State hatcheries to predesignated streams. This program has been judged a major success. It is continuing and will result in miles and miles of additional salmon rivers—supporting hundreds of thousands of returning salmon and countless hours of angler enjoyment—all because one man had an idea and the ability and perseverance to carry it through to fruition.

One of the great stumbling blocks for individuals and groups wishing to do something on their local waters is a lack of money. To counter this problem, many small conservation organizations have evolved successful ways of raising funds. The traditional fund-raising banquet is well known; information on how to run one of these is available from a number of the organizations, including Ducks Unlimited and Trout Unlimited.

Most readers will be familiar with the "walk for water" approach. Here, large numbers of young people are convinced to help. They solicit pledges for each mile walked and at some given time and place great numbers of them take off and see how far they can walk. One small chapter in TU, for example,

Indian Point Plant One on the Hudson River has been responsible for the killing of millions of fish. It has been the subject of action by the attorney general in attempting to collect fines for admitting fish mortality. Probably the greatest danger here, though, as in other plants near striped-bass spawning grounds, is that untold quantities of eggs and larvae have disappeared into the plant and have probably not survived — *since fish don't live better electrically*. Only well-organized group action can prevent such massive destruction by major industrial developments such as this. And even that sometimes does not help. *Scenic Hudson Preservation Conference*

has raised over $3,000 with one of these walks. The sale of pins and bumper stickers can be effective, as may a public raffle to support a particular project or effort. A local business or local chamber of commerce can often be inveigled into contributing to a worthy project. Public funds will hopefully be available through the REAP program again. The Bureau of Outdoor Recreation manages the land and water conservation fund, which also provides funding from the federal government to worthy projects at the local and state levels. Should you wish to pursue this avenue, the first step is to contact the local office of the Bureau of Outdoor Recreation, Department of Interior, and learn the specific ground rules. State fish and game departments often have funds available for stream-improvement projects and they also should be contacted. Don't overlook your local service clubs. Service clubs, after all, are just that. Most of them have within their credo conservation activities and most have established a rather substantial bank account; they can also be a source of ideas and actual assistance.

These are only some of the ways and some of the places where money can be obtained. Few people realize what a tremendous amount of sound advice and financial assistance is available to them just for the asking. Here is a more systematic summary that may be of value.

Advice and Counsel

Government employees, including state and county officials and members of state universities, are, in effect, employed by you the taxpayer. They are, almost without fail, pleased to offer assistance. For general assistance in planning, go directly to the nearest local or regional office, as listed in the white pages of your telephone directory under "United States Government."

1. State Game and Fish Agency
2. Bureau of Sport Fisheries and Wildlife
   United States Department of the Interior
   Washington, D.C.
3. National Marine Fisheries
   United States Department of Commerce
   Washington, D.C.
4. Soil Conservation Service
   United States Department of Agriculture
   Washington, D.C.
5. State universities

If the project you want to undertake is on lands controlled by an agency of the federal government, you must coordinate your plans with that agency. The principal landholding agencies of the United States government are:

1. Bureau of Land Management
2. United States Forest Service

Private organizations that will be of help include:

1. Trout Unlimited
2. The Izaak Walton League
3. The Federation of Fly Fishermen

Financial Support

You will find that the United States government has made substantial monies available to individuals, at the county level, for such conservation projects as tree planting, stream improvement, and pond building. Such public assistance is available from:

1. Soil Conservation Service (under their Rural Environmental Assistance Program)
2. Bureau of Outdoor Recreation (Land and Water Conservation Fund)
3. State Fish and Game agencies

Private monies are often available from the following kinds of groups, if they are effectively approached:

1. Local businesses
2. Local service clubs. Kiwanis, Rotary, Optimists, Lions, and other such groups often have conservation funds available—and need only be sold on the importance of your project.
3. Foundations. (See local listings in the *Foundation Directory,* available at any public library.)
4. Trout Unlimited (normally only through the local chapter or council)
5. The Izaak Walton League
6. Your own fund-raising efforts

Manpower Assistance

This is readily available from:

1. Boy Scouts of America
2. Girl Scouts of America
3. Future Farmers of America
4. Public and private school ecology clubs
5. Service clubs
6. Other local sportsmen's clubs
7. Other local youth groups

Remember, proper planning and preparation are essential. Council with the appropriate agencies and get all the necessary facts before you take any action. Line up your support, organize your efforts, and then implement your plan quickly and effectively.

What can *you* do?

Oh, there are a couple of thousand projects you might undertake. But first, *get involved with an organized group* and become one of the shaking forces in the never-ending fight to reverse the downward trend on our rivers.

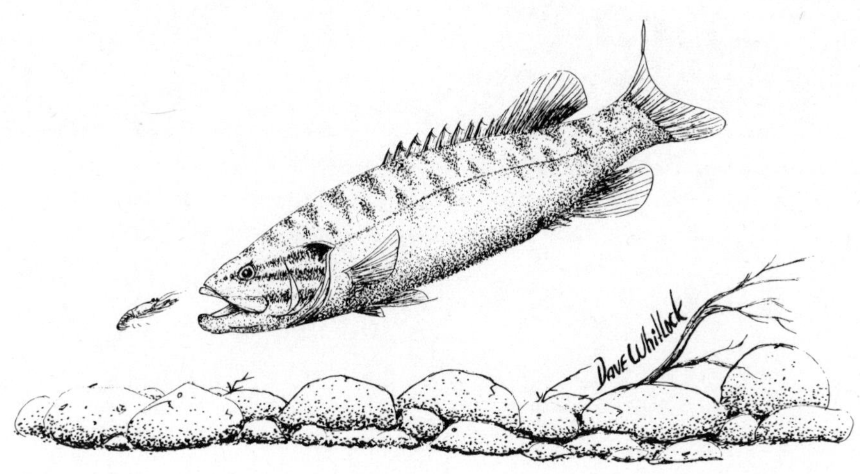

A solitary cane-pole angler makes his last trip to Potatoe Creek in Georgia, just above the advancing dragline. The pleasures of our streams *can be* preserved if those who value them decide that they are important enough for us to join together and save them.

8 Legal Action to Save a Treasure

STANLEY BRYER

" . . . it is a river which is more than an amenity, it is a treasure." [*State of New Jersey vs. State of New York, 283 United States 336 (1931)*]

Such was the sentiment of Mr. Justice Holmes upon delivering the unanimous opinion of the United States Supreme Court on a matter concerning an application to divert water from tributaries of the Delaware River in 1931. He went on to say, referring to the river, that "it offers a necessity of life that must be rationed among those who have power over it." That same decision, rendered as it was before the advent of the recent tide in favor of the protection of the environment, took further specific note of conditions that might "injuriously affect the sanitary conditions of the river," as well as conditions that might ". . . injure the shad fisheries."

It is readily apparent that the protection of the integrity of a free-flowing river was seriously considered in this country at least at the time of this decision. The inroads made by increasing population and development have so compounded the problems that existed forty-two years ago that man can no longer afford the luxury of fouling his resources. The laissez-faire philosophy that flourished through the nineteenth and much of the twentieth century has necessarily given way to a realization of man's responsibility to his fellowman, as well as to his natural resources. The conservationist today is ever more thrown into conflict with the land developer and industry generally, and the necessity of reaching a mutual accommodation has become imperative. Until such accommodation becomes a reality, those who would protect their streams and river resources must resort to those legal tools that are at hand and those that can be developed through the ingenuity of the legal profession.

It is now a principle of administrative law and public policy, as much as it remains a principle of poetry and ecology, that every piece of land and water interacts with every other. So it is slowly coming to be recognized that the individual need not be directly affected, either in person or property, in order to have a sufficient interest to take those legal steps necessary to protect a river or stream.

Before this recognition came about, the conservationist was relegated to the common law principles contained in the law of nuisance. Though a nuisance suit is still a legal tool by which pollution issues may be raised, the proof must show that the individual seeking to maintain the suit is the one who, in fact, suffered injury either to his person or to his property. As a consequence, a downstream riparian owner (one whose property is adjacent to a river or stream) may always sue to enjoin or stop a polluter, or to collect money damages for injuries shown. To establish satisfactory proof of such a nuisance it must be shown that the reasonable enjoyment of one's property has been interfered with, or that the owner's life or health is endangered, or that his physical comfort is disturbed to an injurious extent. The injury must be real, not fanciful, nor such as results in trifling annoyance, inconvenience, or discomfort. The criterion is, in general, the effect of the pollution on an ordinary, reasonable man—that is, a person of ordinary habits and sensibilities.

The law has recognized, since at least the early seventeenth century, that one may not use his property in whatever way or manner that he pleases, without potential liability to his neighbor. However, the law also notes that a person's privilege of making a *reasonable* use of his property for his own benefit, and conducting his own affairs in his own way is no less important than his neighbor's right to use and enjoy his premises. In this country there has evolved a line of cases, going back to the case of Pennsylvania Coal Company vs. Sanderson, which arose in

ber 24, 1972 SUNDAY RECORD

oors: Phil Chase

Winning the conservation battle

d home is becoming as s the second car has ays and produced more tra residence. easonal homes, seldom esidences. Usually the nd new demands ore pressure fro omes taxed pa

or land to be sol aste disposal sy evelopment" co vorable, with the homes and the ce age. ace his own prob tic system in; th siness has the get their foot int supervisors and s

ch money that no land. And with th orm of taxes to se, the eventual hools and the ne es turned over to n to balance out roblems develop sary water and etent receivers ares bankruptcy a

board meetings

able to produce enough water a guaranteed quantity can be obtained by just digging shallow wells near the banks of a stream.

Not all developments produce the aforementioned problems but far too many have

Two problems are faced. One is a scheme llons of water daily fr small Highla

Fireworks Expected At Edgewood Lakes

It is anticipated that Aaron Horowitz, of the engineering form of Eustace & Horowitz of Circleville, will be given hard time on the witness stand when the environmen hearing on Waneta at Liv day morning.

Mr. Horowit testimony whe ton Manor fir first unit of 40 er approval gi

It is expect neys for the environmental seeking to

60 L

Wood, Field an

Building Plan Held Threat to Trout

By NELSON BRYANT

Critics and advocates of a posed 400-home development project in the town of ckland, N. Y., will have opportunity to air the vs at a public hearing next Thursday at 11 A Fire Hall in Living

e hearing was call state's Departme

Recent photo shows payloader dumping gravel into of Croton River bed.

Trout Stream Negle Stirs Up a Sto

By JESSE BRODEY

Gravel mining operations in Westchester's best stream, the Croton River, and failure of the State Resources Commission and Department of Conservation halt them came under fire last week from two more big guns.

In a letter to Gov. Rockefeller, urging him to compel state employes to enforce the laws, Rep. Richard L. Ottinger (D-Pleasantville) charged that the River has been

He asserted agents

of Conservation on the Croton River tion (head

edules earing

which ain 100 uildings, 70 - unit cilities, a ol, tennis s. Snyder Inc., are to develop

water and sewage facilities," Commissioner Diamond said. "We want to determine whether the development will cause unreasonable and unnecessary degradation of the environment."

The development site originally consisted of 99.5 acres along county road 85 in the town of Hunter. Of this, some 13.4 acres has been conveyed to Shanty Hollow Corporation, operators of the Hunter Mountain Ski Bowl, as a new ski lift and trail area.

In plans filed with the department, the developers propose to use as many as four wells as the source of water for the project. After construction of the wells is completed, they would be incorporated in the Water Supply

THE NEW YORK TIMES, SUNDAY, DECEMBER 7, 1969

ists Protest Dredging of Croton Riv

's dredging and related operations."

Mr. Ottaviano has said that dredging operation would prove the river by creating arge lake so that swimming fishing will be carried out more healthful and wholemanner."

hether dredging should be ed in the river has been sue here since last April,

the gravel was being excavated and sold for profit in violation of local zoning laws. The Black Rock area is zoned for residential use only.

Some Cortland residents said that they have followed trucks filled with gravel from Black Rock to the nearby $25-million construction project at the Hawthorne traffic interchange.

In July, several weeks after the corporation had stopped

the third, which charged it impeding the flow of the by building a dike across

William C. Hitt, the Democrat ever elected a pervisor of the Town of landt, charged during the campaign that Croton Realty got "special treat from the Town Board and the board delayed action case until the corporation finished dredging. Mr. Hi

1886, spelling out the right of a downstream property owner who had been injured by the acts of his neighbor. In the Sanderson case the Pennsylvania Coal Company established a colliery on lands owned by it alongside Meadow Brook. As a consequence of the mining operation, water collected and had to be pumped from the mine, and such water ran into Meadow Brook so as to pollute it to the extent that the water coursing through the lands of a downstream neighbor became unfit to drink and caused the death of fish in the stream, as well as the killing of willow trees along its banks, all to the detriment of the Sandersons. While the outcome of this case did not favor the Sanderson cause, the principle of liability for injury to a downstream riparian owner was established.

Unfortunately, during that period of initial industrial growth in this country, and the economic transformation marked by the accumulation and investment of large amounts of capital, the courts showed a tendency to refuse to enjoin or otherwise penalize an upstream owner whose investments were more substantial than those of his downstream neighbor, and they certainly recognized no rights in the individual without a definite property interest. At that time, the fisherman, sportsman, or nature lover, traveling some distance from his home to his favorite stream or forest haunt, and finding it damaged and in process of pollution by some irresponsible individual or corporation, had, and to some extent still has, no recourse at all. No standing was recognized by the courts for the purpose of maintaining legal action, nor was there qualification to appear before a governmental administrative body.

Fortunately, as the shortsightedness of this approach became more apparent, our courts have been inclined, with several notable exceptions, to adjust their sights, particularly where actual injury can be demonstrated by a downstream owner. They are gradually coming to depart from the principles of provincialism and to realize that the rivers and streams of our land are a natural heritage to be enjoyed by all of our citizens.

There has evolved, at least on a federal level, and to a lesser extent within the states themselves, those principles that now recognize the right of any "responsible representative" of the public interest, in an environmental protection matter, to sue in the courts to protect that interest.

Legal Action

In 1965 the first Storm King decision established the standing of conservation groups, under the Federal Power Act, to appear and oppose a planned project, in this case by the Consolidated Edison Company of New York, to construct a power plant on the banks of the Hudson River. It was felt to have a serious potential for environmental damage in that it was "an area of unique beauty and major historical significance." Other federal court decisions quickly followed, culminating in the decision in the Hudson River Expressway cases in New York State, sustaining the right of a representative of the public interest to sue in

The original illustration in a 1963 stockholders' report showing Con Ed's first plan for pumped storage at Storm King Mountain. Later, in 1966, the utility agreed to underground most of the works in an attempt to satisfy environmental objections, not realizing that scenic beauty is only a portion of the environmental battle. Problems still attendant to the project include air pollution from pumping sources, the threats of seepage, leakage of salty and polluted Hudson River water from the upper reservoir—which would endanger local water supplies and upland forests—dangers to the Hudson River fisheries (the plant would draw 9,000,000 gallons of water a minute during the pumping stage) and many other environmental hazards. *Scenic Hudson Preservation Commission*

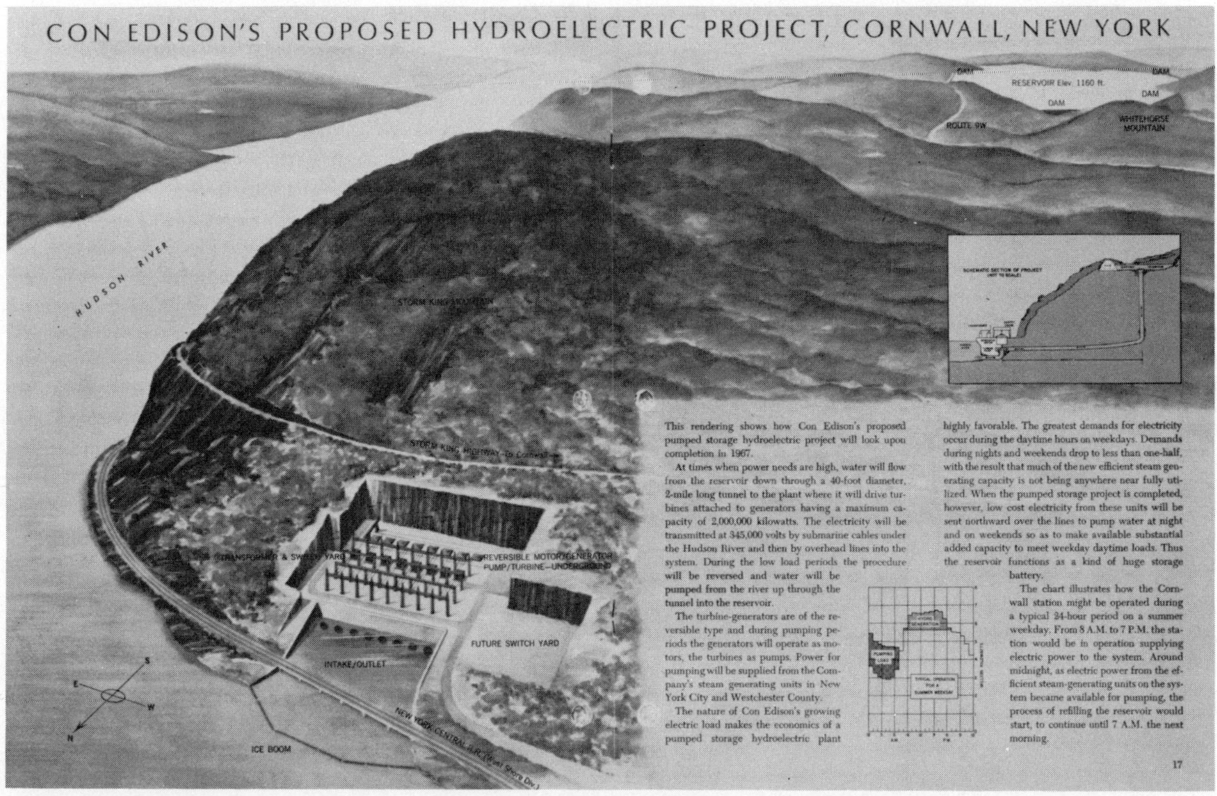

the federal courts to protect that interest in an environmental matter. Involved was the right of various conservation and public interest groups to sue—in this case, successfully—to prevent construction of a proposed superhighway closely paralleling the Hudson River and seriously detracting from the scenic beauty of this historic and lovely resource.

With the state of Michigan leading the way in 1970, a number of states have enacted, and others have under consideration, broad citizens' suit bills to remedy, abate, or enjoin environmental damage, potential and actual. Values are changing and the courts are starting to recognize that a citizen or group should not be denied standing because he or they speak for the public interest, rather than for a private interest. To deny such standing would be to license any violation of law which is on so broad a scale that it injures society as a whole, instead of only a few who may claim special pecuniary damage.

The most valuable weapon, to date, available to combat the polluter of our rivers and streams on a national scale has been the Rivers and Harbors Act of 1888, generally known as the Water Pollution Control of 1888, now somewhat weakened by the Federal Water Act of 1972. This legislation prohibits anyone, individual, corporation, or municipality, from discharging or depositing refuse matter of any kind into the navigable waters of the United States, or their tributaries, without first obtaining a permit to do so. The term "refuse" has been broadly defined by the Supreme Court to include all foreign substances and pollutants, such as solids, chemicals, oils, and other liquid pollutants. The only substances excepted from the general prohibition are those flowing in the runoff from streets, such as storm sewers and effluent flowing from sewers in liquid form.

It is clear that the enforcement of this law seeks and expects the cooperation of our citizenry, and certainly a trout fisherman devoted to his own particular stream should ever be vigilant for potential violations. Violators are subject to criminal prosecution and penalties in the nature of a fine of not more than $2,500 nor less than $500 for each day that the violation is proven to exist. In addition, a citizen or group who informs the Corps of Engineers, or the appropriate office of the United States Attorney, of a violation and gives or provides sufficient information or evidence leading to the imposition of a fine, may

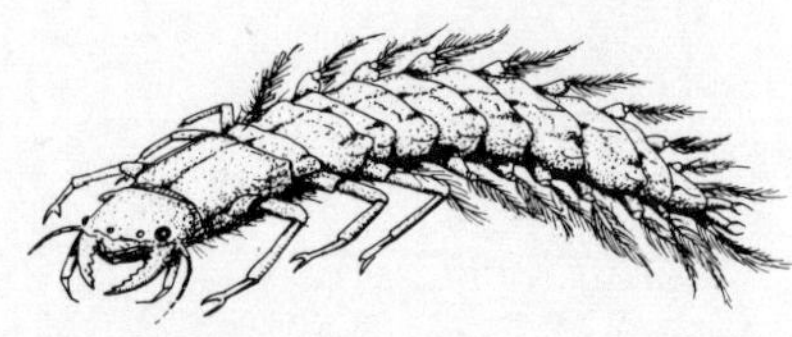

be entitled to as much as one-half of the fine levied. Clearly, we here have an instrument by which the dedicated fisherman and conservationist can perform a service in protecting his beloved waters and, at the same time, find himself compensated for doing so. Where a violation is suspected, evidence should be gathered bearing on the following:

1. the nature of the refuse material discharged;
2. the source and method of discharge;
3. the location, name, and address of the company and person or persons causing or contributing to the discharge;
4. the name of the waterway into which the discharge occurred;
5. each date on which the discharge occurred;
6. the names and addresses of all persons known to the citizen, including himself, who saw or knows about the discharges and could testify about them if necessary;
7. a statement that the discharge is not authorized by Corps permit. If a permit was granted, the statement should set forth facts showing that the alleged violator is not complying with one or more conditions of the permit;
8. if the waterway into which the discharge occurred is not commonly known as "navigable" (such as the Mississippi, Columbia, Delaware, Ohio, Missouri rivers and any of the Great Lakes), or as a tributary to a navigable waterway, the statement should set forth facts to show that the waterway is a navigable waterway or a tributary to a navigable waterway;
9. where possible, photographs should be taken and samples of the pollutant or foreign substance collected in a clean jar which is then sealed. In addition, it would be helpful to collect samples of the water intake in order to show that the refuse material was not in the incoming water but was added by the company when it discharged its effluent. These photographs and samples should be labeled with information showing who took the photograph or sample, where and when, and how and who retained custody of the film or jar.

The Federal Water Pollution Control Act of 1972 created

a new national system of permits for discharges of pollutants into the nation's waters, replacing the 1899 Refuse Act permit program. Under the 1972 law, no discharge of any pollutant is allowed without a permit. This permit program is now administered by the United States Environmental Protection Agency and requires, among other things, that the states establish acceptable effluent standards. More importantly, the act provides for the prosecution of citizens' suits so that individual interested citizens and groups can now take direct action to enforce compliance with federal water pollution requirements and standards. Individuals and groups are now authorized to take court action against anyone violating those standards and, in fact, to file suit against the United States Environmental Protection Agency itself, if it fails to perform any duty mandated by that law. The concerned fisherman and conservationist should find out therefore what environmental standards apply to the rivers and streams in his community and then report violations to the nearest office of the EPA, providing as much of the information outlined above as is possible.

On a local level, the citizen concerned with the preservation of his rivers and streams will generally find that the various state agencies concerned with the environment, such as the New York State Department of Environmental Conservation, have become, in recent years, more responsive to the pleas and complaints of aggrieved individuals and groups. Various states have enacted Stream Protection Laws, and laws and regulations requiring permits before work can be performed in a stream bed or on its banks, or before sewage effluent, treated or untreated, can be discharged into the stream itself. These laws, in most instances, have extended protection to smaller streams and rivers not apt to fall within the aegis of the Federal Refuse Act.

They are, significantly, those streams with which the average trout fisherman will be concerned.

It is here that the average fisherman, outdoorsman, or conservationist can play the most significant role, and, by exercising awareness and vigilance, can prevent a problem from arising and avoid the necessity of resorting to the courts at a later date. By rapidly reporting environmental incursions, fish kills, changes in water temperature, or unusual turbidity to a local conservation officer or agent, and then following up by reinspection at a later date, you will be practicing preventive medicine, which is far better and easier than trying to effect a cure after the damage has been completed.

The local laws and regulations often require that a developer, or other individual seeking a permit, advertise his intention in a local newspaper and provide for the receipt of objections by individuals and groups, not necessarily downstream riparian owners. Here, then, is an opportunity for the concerned sportsman to participate in a quasi-judicial proceeding on a local administrative level. If the opposition is extensive enough, the environmental agency concerned will often order a public hearing on the matter, at which time testimony and statements will be received both in support of the application and in opposition. The trout fisherman and conservationist, working either independently or through a local organization, is afforded a fine opportunity to participate in the democratic process. In order to do so, however, it is not sufficient merely to voice opposition. A strong case in opposition must be made, if necessary by personal study and resort to scientific and technical personnel who, incidentally, are often surprisingly willing to cooperate. If, for instance, an application is submitted for a permit to remove gravel from a stream bed, opposition can be effectively mounted by first ascertaining:

1. the fish population and breeding characteristics of the downstream water;
2. the characteristics of the stream bottom at the proposed work site, to ascertain if there will be an appreciable increase in turbidity with resultant downstream siltation;
3. the effect of siltation on reproduction of fish and aquatic life.

A number of sportsmen's groups and conservation organizations were successful in opposing an application submitted by the Croton River Realty Corporation for a permit to dredge approximately 200,000 cubic yards of valuable construction gravel from the bed and banks of the Croton River in Westchester County in New York State. The opposition demonstrated that there was a valuable fishery within the river as well as at the estuary and that the turbidity caused by the proposed project would create excessive siltation downstream and at the

estuary area with damaging effects on the fish spawning grounds as well as upon aquatic life. It was further shown that the project was neither reasonable nor necessary, nor was it in the public interest. This fight was successfully led by the Theodore Gordon Flyfishers.

If the application is one for a permit to discharge sewage effluent, studies should be made to ascertain:

1. the chemical characteristics and composition of the receiving waters;
2. fish-carrying and breeding-ground characteristics of the downstream waters and bed;
3. rate of stream flow at the point of discharge;
4. waste assimilation capacity of the stream at the point of discharge and downstream therefrom;
5. the nature of the sewage-treatment plant proposed by the developer, in order to be able to predicate the chemical composition of the effluent that will eventually reach the stream;
6. the eutrophication potential of the receiving waters.

While it is clearly not an easy task, it must be remembered that the applicant will be forced to support his position and will probably have to present detailed technical evidence thereon. The conservationist will find that there are lawyers interested in developing and presenting the opposition case in a matter of this sort and, furthermore, that there are individuals possessing the requisite scientific disciplines, such as stream biology, environmental engineering, sanitary engineering, hydrology, and geology, who are willing to make their time and talents available. These people are, in most instances, university connected and possessed of strong concern for the welfare of society.

In many ways, therefore, it is at the administrative level that the fisherman and conservationist will be able to make the most noteworthy contribution. At this level, the rules of evidence as applied in our courts are either disregarded or, if applied at all, are done so in a manner not to preclude an individual who is not legally trained. An attempt should be made by each of us to ascertain what stream-protection laws are in existence in our respective areas, and then to ascertain, in the light of these laws,

Conservation is not diametrically opposed to man's other needs; quite the opposite. There are alternatives: England's Calder Hall Nuclear Power Station employs cooling towers (which emit steam, not smoke) instead of river water. The total capacity of the station is 198,000 electrical kilowatts. Effective early legal action can often persuade large corporations to include conservation among their list of priorities.

how opposition to the pollution and destruction of our beloved waters can be mounted most effectively.

A case in point is one that arose as the result of an application for a permit submitted to the Department of Environmental Conservation of the State of New York. The application sought permission to draw up to 280,000 gallons of water per day from the Schoharie Creek in Greene County in New York State, for the purpose of supplying a proposed condominium-type housing development at Hunter, New York. Permission was further sought to construct a sewage outfall in the creek.

The opposition, once again led by TGF with assistance from Trout Unlimited, successfully demonstrated that the effects of the withdrawal of the amount of water indicated above would be catastrophic, considering the stream characteristics of the Schoharie. Those who have fished it extensively know it to be extremely low and warm in midsummer and low again in midwinter, the periods when the water demand at the proposed development site would probably be the greatest. It was further shown that the developer failed to consider adequately the effect of this project on the ecological balance existing in the Schoharie at the point of discharge and downstream. No significant attempt had been made to study alternate sources of water supply. As a result of the timely action taken, the developer was forced to withdraw his application and to commence in-depth environmental studies bearing on the questions of water supply and sewage disposal.

Perhaps, we should all become cognizant of that which may well be one of the conservationist's strongest allies available in the continuing battle, that is the constantly growing element of public awareness of and repugnance at the frittering away of our valuable natural heritage. We should not hesitate to solicit the assistance of the press, because there is little doubt that exposure to public scrutiny is the greatest leveler when we find ourselves pitted against industrial and development interests with vastly greater monetary resources. The press is in the news business. The environment is news. Responsible citizens' organizations in the environmental field should and can be part of the newsbeat. The press is generally sympathetic to the environmental cause, as most people are, and the press can do something about it. It can use its editorial columns to support the conservation position, and it can cover significant environmental developments in its news columns. The local newspapers were of valuable assistance in both the Croton River and the Schoharie Creek battles mentioned above. The attitude of the public generally can be well gauged by the pressure they apply to their elected public representatives, which then becomes evident in the legislation enacted. Here I have specific reference to the endorsement of an Environmental Bond Issue, such as was recently voted and approved overwhelmingly by the citizens of the state of New York, as well as to the pending Citizen Suit legislation in New York State and elsewhere. Also

A commercially logged slope just above one of the major steelhead streams in western Washington. This kind of logging is legal, despite its consequence, which can easily be imagined. Where current laws are inadequate, or ineffective, they must be changed.

in point is the comprehensive Land Use Planning legislation recently enacted in the state of Vermont.

It is and must be, however, the willingness of individuals and groups, on a local level, to engage in administrative proceedings, where possible, and in appropriate legal proceedings through competent counsel, where necessary, that will carry the day for those of us concerned with the protection of our rivers and streams. Nor should the average citizen feel he has no recourse in the proper forum against the elected or appointed public official or administrator who, either through nonfeasance or malfeasance, fails to carry out the mandate of his office or of the people whom he serves. Most jurisdictions have established the necessary machinery for legal review of administrative decisions, and the courts have shown a willingness to entertain the same, where it can be demonstrated that actions were taken or neglected in derogation of authority mandated. Consequently, the sportsman and conservationist, in addition to considering ways in which he can become a participant, should also be ready to scrutinize the decisions made by his elected and appointed officials.

Every fisherman and every individual who derives inner satisfaction from the sight, sound, and presence of clear, clean water, has a responsibility which cannot and must not be ignored. Our water resources are not limitless and, generally, when lost, are lost forever. Those of us interested in stream preservation are, fortunately, many in number and well dispersed so that, with the proper exercise of vigilance, none of our resources that are worth saving, in any section of our country, should be abandoned to industry, development, or just plain carelessness, without inquiry and a proper fight mounted if deemed necessary. There are, throughout our country, conservation organizations willing to lend assistance and advice and, if necessary, to take up the cudgels of battle. In addition, there are knowledgeable lawyers and associations of lawyers generally concerned with environmental law, whose advice is available when needed. Those organizations that are concerned and are particularly well qualified to assist the individual or group in need of help in protecting the integrity of a valuable river or stream are the Sierra Club through its many regional offices in the United States, the Theodore Gordon Flyfishers, a regional organization having its principal office in New York City, Trout

Unlimited, a national organization with regional offices well dispersed throughout the country, as well as the Environmental Defense Fund and the Natural Resources Defense Council, Inc., a public interest law firm with offices at Washington, D.C., and in New York City. (See the back of this book for addresses.) There has been no attempt to list all possible organizations that could or would take an active interest but merely those which by their nature are constituted to be active in the field and which have staffs of attorneys available.

Renowned humorist James Thurber once observed that "Man is traveling too fast for a world that is round. Soon he will catch up with himself and man will never know what hit man from behind was man." Perhaps those of us with sufficient interest in the preservation of our natural heritage can assist our fellowman to see the trail of waste and destruction that he is leaving behind and so help him avoid that final catastrophic collision. Then legal action will not be necessary.

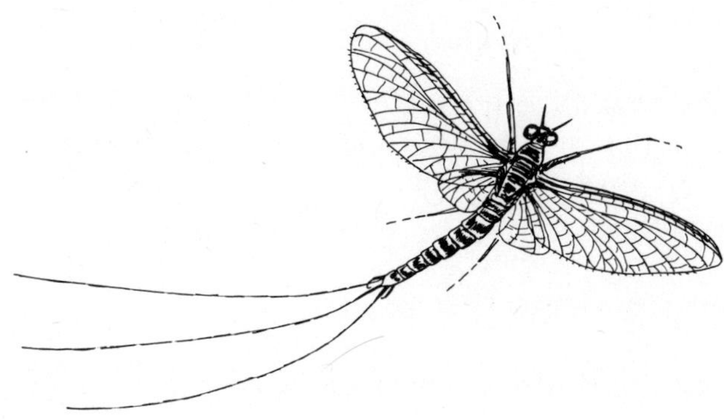

9

An Overview—Reflections and Prospects

ALVIN R. GROVE, JR.

Our environment problems are man-made;
the solutions must be man-made as well.
SENATOR GAYLORD NELSON

In many ways, the qualities that make water a suitable habitat for trout and for other species of fish, for other vertebrates, and many invertebrates are basic and elementary. In other respects, these qualities are so complicated that much of the knowledge necessary to understand all of the ramifications has not yet been accumulated.

Much of what is true in nature—the basic needs, limitations, expansion and invasion, success, failure, variation, and other aspects of the continued existence of living things—can best be understood if we remember that we, too, are animals and, like the trout or the mayfly, demand and rely on a hospitable environment for our existence.

In some ways, it is easier for man to understand the environment as a place where life exists, if we look at those things with which we are most familiar and, incidentally, apply the yardstick of measurement to the living thing we may know best—ourselves.

Two recognizable, basic biological phenomena are related to all species: one is the continued survival of the individual, the other the continuance of the species as a group of like individuals that interbreed and constitute a population. An individual attempts to live—a phenomenon equally true of man, a fox, a bird, a trout, or a stone fly. It is a fact that these species go about it in different ways, as a result of the level of development, but this matters little to the basic theme of survival. Obviously, many factors determine the chance of survival, based on the comparative needs or requirements of different species.

In general, it is assumed by biologists that whatever exists at the moment is the end of a long chain of events. The first birds did not look exactly like the ones we know, the first fish did not look like the ones we are acquainted with, and the first so-called man did not look quite as we do. Gradual, subtle changes have taken place over long periods of time, and, in each instance, it is assumed that the particular individuals that survived were those able to adapt to meet the new demands of the environment. Actually, the relationship is probably more complicated than this, with the continued survival of a species dependent on sufficient variation within the many individuals of its kind. Among all members of a population, e.g., tall, short, fat, thin, scaled, not scaled, fins, no fins, some individuals could and did, in fact, continue to exist, flourish, and reproduce within the changing environment. Those individuals whose characteristics make their survival possible are said to be adapted to the environment. There is a subtle but significant difference between this interpretation and another which assumes, for example, that trout developed or grew fins so they could swim.

In fish, we are aware of the fact that suckers might live where trout cannot, that even among species of trout some can withstand higher water temperatures than others, and even within a species some, apparently, can withstand higher concentrations of acid or nitrogen than others. Most biologists believe that the variation existing among individuals of the same species or between one species and another is a major key to the continuance of a kind of animal or plant. Although not the only factor in operation, it is possible that the so-called gene pool that contributes the greatest variation with the largest number of possible recombinations produces offspring with the best chance to adapt to changes in the environment.

It is clear then—at least, in a general, broad-based interpretation—that fish require the oxygen of the atmosphere to exist (just as we do), but that some species can survive with a little less and some do better with a little more. There is a limit, however, and if no oxygen exists then the fish cannot survive. If we decide that suckers require less oxygen than trout, expressed as parts per million, then we might conclude that we will get trout and not suckers if we can provide more oxygen in solution in the stream, but this is not necessarily true. Water must be free of certain contaminants, and, obviously, temperature also influences the distribution of kinds of fish. Within the ecosystem any single factor or several factors in combination play different roles as major or minor determiners.

All life—plant or animal—exists in a relatively thin shell around the surface of the earth. In this area certain essential elements—like air and water—are almost universally present. Air, especially the oxygen of the air, and water are basic needs of living things. In a few instances, where oxygen might not be needed and be, indeed, detrimental to an organism, a specific microclimate must be established.

The planet earth derives its energy from the sun. Much of what we use comes each day, but some that reached the earth millions of years ago has been converted into deposits of fossil fuels, which we now mine and burn as coal and oil.

Three major elements of the environment are omnipresent and in most respects uncontrollable by us or any other living thing—water, air, and energy (heat). For most living creatures to survive within an ecosystem, the system must contain oxygen, water, and a suitable temperature. The tolerance or adaptability of a species will permit it to exist within narrow limits, i.e., to tolerate minor variations in the oxygen content of the air or, in the case of trout and other fish, in the oxygen dissolved in the water. Trout will do very well at seven parts per million of oxygen, they will likely do better at ten parts per million, and they will manage to get along for short periods of time at five parts per million. The atmosphere, from which comes the oxygen that is dissolved in the water, contains other gases as well, and these, too, are soluble in water. As a matter of fact, the nitrogen of the atmosphere, which is much more abundant than the oxygen, is also dissolved in the water and, in some instances, is in such large quantities that it becomes toxic and consequently limiting to the distribution of fish, even causing their death. It is known, too, that certain impurities in air not

normally found there but provided by man's industries will also dissolve in water.

Temperature is a third basic criterion to consider at this level of examination of the environmental necessities. Not only does temperature (heat) directly affect the trout, but it also has an indirect effect on other phenomena. We say, for instance, that trout are cold-water fish (cold-blooded as well) and demand temperatures under 70° F. This does not mean that any or every temperature under 70° F is equally as good for the trout, nor does it necessarily mean that every temperature over 70° F will cause the immediate death of trout. But 70° F is essentially the upper limit or maximum temperature that trout will tolerate. The same adaptability referred to before in relation to the amount of oxygen dissolved in the stream water is reflected here as well.

When the temperature is low, e.g., 40° F, the trout's activity is reduced. The fish is less active and so are its metabolic processes. The metabolic processes are all of those things happening within the fish that keep it alive, such as digestion, respiration, and enzyme activity. As the temperature increases, the activity increases so that trout feed more readily—they rise to the fly, they seek out mayfly nymphs. When the temperature rises above a comfortable point, the trout become exhausted, sluggish, and, in fact, may either have to move to a place where the temperature is lower or die.

Because temperature affects chemical reactions and the behavior of such gases as oxygen, it is evident that more oxygen is required if an increase in temperature speeds up respiration in the trout. But if the activity of oxygen itself is speeded up with an increase in temperature, then it tends to pass out of the water as a gas rather than being dissolved in the water. In other words, the high temperature not only makes the demand for oxygen greater but simultaneously lessens the amount of it available in solution. Temperature, so important to the fish, is equally important to fish food.

You may well ask, "What is this all about? This guy is crazy for details but what do I do with them?"

An understanding, or, perhaps, appreciation, of a rather simplistic explanation of some basic facts involving a part of the ecosystem is absolutely necessary to proper decision making, and it will provide most of the explanations to the problems encountered in fisheries biology, stream pollution, stream improvement, better fishing, and greater productivity.

If we assume that your favorite trout stream has a very low population of trout and you want more trout to fish over, it might be suggested that you put in a wing-wall deflector to create a swift run or a low dam to create a pool where you can fish. This may be a nice project to work on. You and your buddies get your feet wet, and you may feel noble about having sacrificed several hours of angling for the good of the trout. The structure is a masterpiece of engineering, the logs are long and straight, and they are placed well back into the bank. In fact, a nice run is created. The pool you had visualized is formed, and you feel good when you fish there. But, alas, you still catch a chub or a sucker because the structure did not solve the problem—probably because the real problem had never been identified in the beginning. Temperature was the problem, and, in spite of the new structures, no significant change could take place because they had nothing to do with solving the problem.

There are some, yes, many streams that, during the warmer part of the season, have hatches of very small mayflies. You are told, and you are ready to agree, that nothing larger than a Number 22 will take a rising trout. Some stupid fly-fisherman who hasn't gotten the word nonchalantly fishes the water during the middle of the day with a Number 10 all-hackle fly and, in the small stretches of white water, takes trout after trout. But the explanation is simple, and it rests on the fact that, as the temperature of the summer has increased, the parts-per-million oxygen content has decreased. The trout, hunting a comfortable location, move to the white water where oxygen is being added to the water as it splashes into whitecaps. The trout are there because the oxygen is there, and they will feed if something that looks like food makes an appearance.

We are acquainted with larger waters that are warm in the summer, but their feeder streams are cooler, even to their

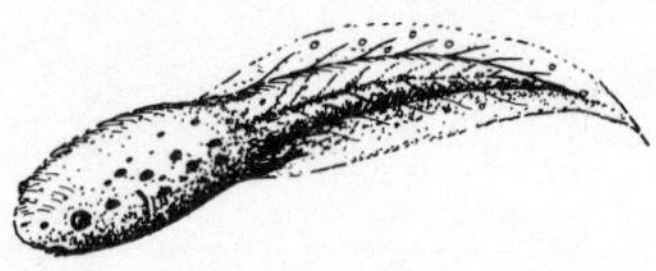

mouths. During the daylight hours, when the sun is high, the fish congregate at the mouths of the feeders. Usually they are restless and not very comfortable; normally, they will refuse to eat. After dark, when the sun sets and the water temperature of the larger receiving stream decreases, the trout gradually disperse into it and begin to feed in the riffles or in other places where food is available. Night fishing is excellent. But for the uninitiated, who continue to fish these same spots the next morning after the sun comes up and the water temperature increases, there is only, at the most, a harvest of rough fish.

When you read about nitrogen supersaturation on such rivers as the Columbia and Snake, it must be remembered that the gases of the atmosphere are soluble in water. Nitrogen is the most abundant gas in the air, constituting about 78 percent of the total, with oxygen running a poor second at about 20 percent, and other so-called rare gases making up the remainder. Ordinarily, the pressure of gases dissolved in water is no greater than that of the atmosphere because gases in solution escape until the pressures are equal or in equilibrium. In shallow streams dissolved gases tend to escape rapidly as the water splashes over rocks and small falls until equilibrium is established. In the case of nitrogen supersaturation on the Snake River, large amounts of water are discharged through the spillways and plunge into deep pools below the dam, forcing entrapped air into solution. Because nitrogen is the most abundant of the dissolved gases in the air, it contributes the greatest amount of pressure. The high nitrogen content may produce minor signs of distress, such as gas-bubble disease, or many fish may die with symptoms similar to the bends.

What, in fact, we are trying to say is that a relatively simple understanding of the basic demands on the environment for the survival of a trout, other fish, or fish food is absolutely necessary. Not all problems are man-made although a great many are. There are natural springs with high concentrations of dissolved nitrogen, and many limestone springs cannot be used as a source of hatchery water until the nitrogen has been liberated. Temperature of water used in hatcheries, especially in the winter, is often low enough to prevent, or, at least, delay, growth during that time of the year, making the cost per pound of fish produced exorbitantly high. Frequently, these hatcheries are aban-

A large limestone spring out of which flows a stream in central Pennsylvania. The water in the spring is high in dissolved nitrogen, and trout that are held there in wire baskets develop symptoms similar to the bends and soon die.

doned or some system of heating the water is incorporated into the hatchery construction—not a small price to pay for the pounds of fish produced.

It would be desirable to sort the environmental factors into little packages and to deal with each in turn. It would be nice if each factor, when changed to whatever the best condition of input might be, would automatically turn up with the best population of trout, especially those that would readily feed on fly or bait, that would fight like a champion, and if eaten would have the most delicate aroma and the most pleasing taste. Unfortunately, all of this is a dream and beyond reality. It is possible, within rather narrow limits and in an environment

This stream, draining a mountain meadow, is naturally acid from decaying plant materials including a variety of mosses, cranberry, and huckleberry. No trout can live in this water.

where trout or other fish already live, to manage the population for the best sport. But to start from scratch to create a trout stream could easily be beyond our knowledge and ability. In some rare instances, where someone has gone out on a limb and estimated the cost of doing this, the price tag has been about $1 million per mile.

The action and interaction of the factors of the environment with the biological entity under discussion is called ecology. Within the system of action, interaction, and reaction, an ecosystem is most complicated and at times defies description or understanding. Changes within an ecosystem are most readily understood when a single factor is found to be the total causative agent of the problem.

If high temperature alone prevents trout from inhabiting a stream, then the lowering of the temperature may afford a simple and effective answer. It may be easy to reduce the surface area, which absorbs heat, but if the high temperature has also destroyed the insect life, thus removing the trout food, the problem is complicated beyond an easy solution, and the development of shade will not in itself produce trout water—especially if we are talking about a population that has some capability of being stream-bred and self-sustaining.

Earlier in this chapter we referred to two broad, general biological concepts—one having to do with the survival of the individual and the second with the survival of the kind, the species. The significant point is that not only are such factors as water quality, atmosphere (oxygen), and temperature important to the individual fish as it lives from day to day, but these factors and many others are also important to the reproduction of the species. As a matter of fact, it is easy to understand that not only water, oxygen, and a suitable temperature must be a part of the life of both the individual and a reproducing population, but many other requirements are also necessary. Some of these are chemical and others are physical features of the habitat.

Briefly, we might consider, largely by suggestion, some of the physical features of the habitat that are prerequisite and must

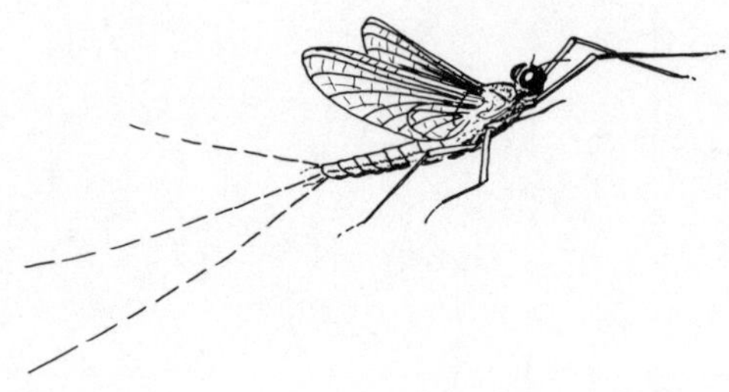

exist if the trout or other aquatic life is to be sustained from day to day, week to week, or from season to season. The riffles of streams are the most productive areas in terms of food: aquatic insects and other invertebrates are most often found here. In addition to these relatively shallow areas, there are deeper pools, which are not very productive in terms of food, but do provide places for the fish to hide, perhaps, to feed, and are valuable in high-water periods and for overwintering.

Trout in shallow water may fall prey to predators, but in deep pools, especially if some additional protection is offered, such as an overhanging bank, large rocks, or dead tree branches, the fish are relatively safe. Stream-bred trout are much more aware of the presence of predators and move much more quickly to places of safety than many stocked trout. Native fish seem to be more independent of each other and do not school like the hatchery product so that predation, although one fish may be taken, is not likely to result in a number of trout being killed or lost at the same time.

The pool-riffle ratio is important to a trout stream and will determine to a considerable extent the possible carrying capacity of the stream. Food and a place to live are obviously as important to the trout as they are to man. Places to loaf and rest, places to feed, places to hide from predators (including man), suitable temperature, in addition to the chemical properties of water, oxygen, and other gases, are all required. It is difficult, if not impossible, to determine which of all of the factors necessary for survival of the trout are most important. Probably, there is no single factor that can be identified when it is recalled that all conditions vary through a small acceptable range when considered in light of the biological limitations existing within the species—trout or trout food.

Acceptable temperature, pH, and oxygen may permit trout to inhabit a stream or lake even though food may be a limiting factor, and the trout may grow slowly, often with large heads and thin tapering bodies. The few conditions of the environment mentioned above could be arranged in several sequences, and each condition might permit the trout to exist but limit its development, its activities, its maturation, and no doubt the population supportable under the adverse condition of a single limiting factor.

Some additional conditions of the environment are prerequisite to spawning—that part of the life cycle of a trout that makes it possible for the species to be reproduced as compared with the comforts and survival of the individual. There must be suitable spawning areas of gravel, the correct size and depth, with an adequate supply of oxygen, which is sometimes provided by the presence of an underground spring in the stream bed. Acceptable temperatures are also necessary, but, within the limits of tolerance of the eggs, the effects of temperature may relate more to the time or period of hatching rather than to their hatching or not hatching.

The words *suitable spawning area* appear to be clear and precise enough for general use but, in fact, the conditions inferred may be many in addition to the few factors mentioned. Suitable and tolerant temperatures, oxygen, size of gravel, the ability of the female to make the redd, the general absence of fine materials that would prevent the eggs from securing the necessary oxygen after they were deposited, sufficient depth to reduce predation are only a few of the necessary conditions.

The angler and, yes, the biologist and the fisheries manager must constantly be impressed by the combination of chemical and physical conditions of the environment that must match the biological demands of each species. The relatively narrow, acceptable limits are compensated for, in small part, by the adaptability of the species. It is a wonderful thing that trout streams and trout exist. The evolution of the stream, of the trout, of the trout's food, and all the other factors that could be mentioned have come about over thousands of years and are the temporary end product of evolution. It is somewhat preposterous of us to think we can change much of this and yet simultaneously wonderful that we can change any of it.

It is evident that the most important job ahead is to prevent the loss of the resource. By resource is meant that combination of factors that make a trout stream possible. Most men who think about the environment are just beginning to realize the fragility of a trout stream, a bass stream, or even a redeye fishery. Those irresponsible people who argue that anything they do resulting in destruction will be repaired by nature in

just a short time are certainly ignorant of the facts, and we suspect they care little whether nature can or cannot repair the damage.

There is ample evidence that certain temporary damages can be corrected, but even in these instances we are not certain of the time required. From surveys made in some streams to determine the productivity of stream sections it is now known that gross disturbance in the stream bed nearly a century ago is still reflected in an 80 percent or greater reduction in productivity. *The person who channelizes a stream, who causes a gross disturbance of the stream bed, or who diverts the stream into a new channel is performing major surgery that may never heal, or if it does, certainly, not within the lifetime of the offender, or his children, and, perhaps, not within the lifetime of his grandchildren.*

What can we do? What are we talking about when we refer to streamside conservation?

I suspect that first of all we are talking about how to save or to improve known trout water; we are *not* talking about creating all the chemical and physical conditions necessary to make a trout stream. In other words, if the problem is small enough and we are intelligent enough to recognize it, we might be able to improve or to correct the problem and to make recovery or improvement possible.

We may make all of this sound too impossible—and that would be wrong. Much, in fact, has already been done to assist some biological species as is evidenced in stream reclamation, introduction of species into waters that formerly could not support them, creation of artificial or new spawning areas, improving, in some instances, the pool-riffle ratio of streams resulting in better productivity, as well as other specific and no less startling improvements. As more of us become interested in such things, the greater the possibility of these occurring. Still most important is our determination to prevent the initial destruction of the environment. This is much cheaper, takes less time, and capitalizes on what nature has been busy doing for at least the past 10,000 years.

There is no real intention that discussion in this chapter be related only to trout, or even restricted to trout streams. It is a little easier to recognize the streamside problems related to our waters and their inhabitants if we limit the concept to smaller

Canyon Creek, Oregon, 1972. The boulder emplacements have helped to create stable pools and gravel bars that are inhabited by trout and are being used for spawning. Cottonwood and willow form dense stands along the bank. *Photo from* Trout, *courtesy of Oregon Office of Federal Highway Administration*

streams. It obviously does not follow that all small streams are trout streams but this is sometimes so. Certainly, those smaller streams that in many parts of the country are excellent bass water should not be excluded from the attention of willing helpers who want to make things better.

But there are many larger streams, rivers mostly rather than creeks, runs, or rills, that are neither easy to understand nor easy to change with a low dam, a single- or double-wing wall, or any of those other nicely pictured devices in so-called stream-improvement manuals. It is the opinion of this writer that much of what must be done to preserve the clean water we have—the fishable water—or to correct many of the mistakes that have already been made through ignorance and stupidity must, in fact, take place in the legislature of each state and in the United States Congress. The main concern of conservationists must be to help in the preparation of legislation that accomplishes those protective, corrective measures that must be taken to *preserve* the environment—that narrow band of life-supporting biosphere surrounding this planet.

Much of the Susquehanna River in Pennsylvania could be excellent bass and walleye water; but the primary problem of this river is acid mine drainage from the hard- and soft-coal fields. The West Branch of this river flows through much of the soft-coal region, receiving the drainage from thousands of square miles of land that has been upset and laid bare through open-pit mining practices. Parts of the North Branch drain the famous anthracite region of northeast Pennsylvania. Hundreds of miles of these branches as well as sizable tributaries support essentially no biological life. There is neither fish nor fish food, and, certainly, there is nothing that can be done with stream devices, by the planting of shade trees or grasses along the banks, or with protective structures within the streams themselves. The solution of this tremendous problem must stem from the halls of Congress with the expenditure of hundreds of millions of dollars to backfill, seal, and reclaim the land from which the acid water originates.

We should learn a lesson from this terrible loss of environment, a lesson that sometimes seems never to be learned no matter how many times the story is told. Mining laws in some states have been strengthened to permit mining without pollution, insisting that the land be leveled to at least the original contour and seeded with trees or grasses. But in many places even this minimal requirement is not being demanded of those who take the resources from the ground. In this time of shortage of fossil fuels and other materials from which we must derive our energy, the necessary legislation should be passed, the proper regulations and rules promulgated, and an inspection force should be ever present. Too few anglers involve themselves in this most important part of the conservation picture today.

We have said a little about acid mine drainage, but there are at least several, and perhaps many, basic problems of land use that are the seat of most of the present trouble and will, in retrospect, be judged to have been the great deficiencies of the seventies and eighties. Land-use planning, which includes the permanent or the temporary removal of its riches, should be the first concern of most of us. The devastation that results from the building of highways, irresponsible timbering, farming, irrigation, the drainage of swamps and other natural aquifers, the mining of clay, the dredging of our streams for sand and gravel, and other activities are the basic causative agents of the present, continued destruction of our water resources. To cope with these, we humbly suggest that the best thing to do is to put the boots and the willow cuttings down and pick up the lawbook, the ballot, and the pen. This must be the course of the future if we are to influence significantly what is done, what is lost, and what is saved.

This approach seems long and hard, and many have already gone this route, spending a lifetime at it, only to come away empty-handed. It is a discouraging picture, but today, perhaps, more so than at any time since the white man colonized America, many are aware of what has happened, and is still happening, to our environment.

We must band together, joining active conservation groups that operate at all levels. Earlier, it was pointed out that there are forty-three million anglers in this country. If each of us would do one thing each day, miracles might be wrought.

Moreover, there remains the little problem of doing something about what we still have. For instance, there is the trout stream or the small bass stream that is already supporting fish and fish food that undoubtedly could be improved, or water that will hold trout for six months of the year but loses the fish when the oxygen goes down to zero parts per million or the temperature climbs to 85° F might be improved and changed into a stream that could support trout year-round, and, with a little luck and common sense added to hard work, there might be established a population of trout that would ultimately fill the water to capacity with stream-bred fish. These possibilities are not to be overlooked, and we should be working on them but not with eyes closed to the clear-cutting at the headwaters that will, in spite of everything we might do someplace downstream, still result in rising temperatures and deposits of sediment feet deep in some places.

In many quarters today those interested in water pollution contend that sedimentation or siltation is the worst of the pollutants. It seems hard to come to this conclusion when we think about the many noxious materials added to our waterways—the toxic ions, harmful chemicals, organic waste products from treatment plants, the results of wood cutting and debarking as well as those chemicals that get into our watercourses from the fabricating plants of paper, hardboard, and other wood products. But in spite of such wastes, the sedimentation that comes from our land not only represents the depletion of the topsoil but it also covers every natural spawning area, irritates the gills of every fish, and smothers every living insect or other invertebrate in the stream.

The removal of such sediment from the stream is a terrific job. As a matter of fact, we know of no effective way to do the job although we have read about, and seen pictures of, the so-called riffle sifters used in some streams in California. The general absence of information about this technique in recent years makes us wonder whether it, too, has failed to produce the desired results.

In some areas, the cooler waters of deeper lakes no longer provide refuge for trout during periods of high stream temperatures because of the excessive rate of sedimentation. In some streams clean gravel bottoms that provided spawning places for trout are now covered with several feet of mud deposited from soils disturbed by farming, livestock feedlots, highway construction, clearing of relatively large areas of land for shopping centers, and a myriad of other activities.

In one such stream close to my home in Pennsylvania, several feet of silt have for a number of years provided the place of anchorage for a tremendous crop of water weeds that had not grown there before. In some seasons of the year the weed growth is so lush that it is difficult to see the surface of the water. When these weeds begin to die, toward late summer and early fall, their decomposition (or oxidation) uses all of the available oxygen in the water. During the daylight hours when photosynthesis is taking place, the fish—if there are any—manage to get some oxygen, but when the sun sets, the plants continue to consume it, and by three or four o'clock each morning the oxygen content is actually zero parts per million. Weeds from upstream, which have broken from their moorings and are partly decayed, float downstream to repeat their oxygen-grabbing process several miles below so that the destructive removal of the oxygen is not necessarily limited to the place of the original growth.

For the sake of illustration, I have taken the liberty of sorting out sediment, weeds, and oxygen consumption, but the problem is really not that simple. Thirty years ago, a single sewage treatment plant dumped its partially digested wastes into this stream. Usually, the remaining oxygen demand of this material was satisfied before it moved very far downstream. Because only a short distance of fishing water was involved, the loss was not considered to be serious. Today, there are five sewage treatment plants dumping their effluent into this same stream. To be sure, each is a better plant than the one of thirty years ago. The treatment is at least 85 percent efficient, and, at times, even more efficient than that. But the total volume of material that is spewed into the stream, demanding oxygen for its continued oxidation, is many times what it was years ago. There are more people, better central collecting and treatment stations, better and more efficient plants, and, finally, we have pollution on a grand scale.

This stream contains few, if any, trout today except some that are stocked during the early months of the season. Those that are stocked are, of course, doomed to death either at the hands of the angler or by starvation. If, by chance, they last too long to fall prey to either of these fates they must inevitably die from lack of oxygen. These problems will not be solved with the planting of willow shoots and dogwoods or with the building of wing walls and jack dams or by putting large rocks into the stream to provide cover.

But these problems could be solved by requiring that all wastes entering the stream exert no additional oxygen demand on the stream. Each treatment plant would have to have tertiary treatment to handle the excessive phosphates, nitrates, and any other specific material which proved to be toxic or nutritive. The silt would have to be removed. It might be taken out by hand, by a riffle sifter, or hydraulically. Certainly, these are rather drastic measures and ordinarily would be destructive to the aquatic insects and other food available, but the destruction of these has already been completed. Some fencing of cattle, the removal of barnyards from the creek bed, proper protection to prevent new siltation from highways yet to be constructed and houses yet to be built would all save the stream for the future.

Having accomplished this much of the recovery, fish could be stocked. Hopefully, aquatic insects from the area would, in some period of time, repopulate the stream. This process might be speeded up by the introduction of insects from other streams. Although little has been done with this technique in this country, it has been pursued in England and other European countries for years.

With luck, the stream could be restored. The volume of water is greater today than it was thirty years ago because many wells have been drilled to supply water for many families, and this is now all collected and directed eventually into this stream. Temperatures are still satisfactory. The stream still flows through a more or less rural area, and although many homes have been built, they are not close enough to the stream bank to affect the water temperature. Furthermore, this is in limestone country, and large natural springs feed the stream millions of gallons of water a day at temperatures that almost guarantee that the trout can live happily in it.

We suspect that nothing will happen in the foreseeable future. Someday, it might be decided that the value of this stream, about seventeen miles of it, is so great that we cannot afford to delay any longer. If we can spend a million or more dollars a mile for new highways, we might well expend the money to correct the wrongs done to this once outstanding trout stream.

Needless to say, there have been extensive biological studies performed on this creek. The problems are rather well understood and much data on BOD, COD, oxygen present, numbers and kinds of fish remaining, aquatic insects still present, and other facts of importance have been gathered. In 1972, the exceptionally high water from hurricane Agnes washed the silt from some places in the stream, but even forces such as this failed to move much of it downstream. In fact, if all the silt were to be moved downstream and into the receiving stream, there would be created at another location much of the same problem.

Much of what is done in the way of so-called stream improvement is worthwhile even though the problem was only eyeballed and the solution determined by amateurs. Certainly, the riprapping of a bank that is being washed into the stream and is covering the spawning beds and the niches where the aquatic insects live is in itself worthwhile even though this may not be the single factor that restores the stream or improves the fishing. By preventing loss of banks, the stream is confined to a smaller cross section, and less surface is exposed to absorb heat. The smaller cross section of the stream increases the rate of flow, and this in itself helps to scour the bottom clean of fine materials, providing a better habitat for fish food and potential fish spawning.

The creation of hiding and protected places for fish cuts down on predation and helps to stabilize the carrying capacity of the stream at somewhere near its upper limits rather than at some lower level. The planting of banks, especially with grasses, not only helps to stabilize them but also produces an environment in which many insects might live and periodically fall into the stream to add to the food supply. The fencing of cattle from the stream bank is always worthwhile and, in certain instances, may be the only corrective action needed to restore a natural spawning area. If the only limiting factor for trout occupancy is a temporary rise in water temperature to slightly above tolerable limits, the planting of trees and grasses that provide shade might well be the single solution to the single problem. Where gradients are steep and many shallow riffles exist, which usually are good food-producing areas, but there are no pools in which fish can loaf or rest, then the creation of pools might well supply this missing element of the stream's environment. In some cases, springs that feed streams or even form streams contain large quantities of dissolved nitrogen. If this water is agitated or stirred over a low dam of rocks or logs much of the nitrogen will escape as gas, making the water much more suitable for trout habitat. Occasionally, such water is collected and sprayed mechanically. In other instances, it is dropped from the top of standpipelike arrangements to fall over a series of pools of increasing size as it loses its nitrogen. Small amounts of acidity may be removed by the addition of limestone beds or by passing the stream through boxes or containers of lime or limestone. Generally, this is an expensive and very demanding process, and

Big Roche-a-Cir Creek, Waushara County, Wisconsin. Meadow section before current deflectors and bank "hides" were installed. The deepest place in the stream within view of the photo is a small twelve-inch pocket. Most of the stream, at mid-channel, is only about six to ten inches deep. *Photo by Ray J. White*

The meadow section of Big Roche-a-Cir Creek about two years after current deflectors and bank "hides" were installed. Many places in the picture are now eighteen to twenty-four inches deep; with overhanging cover. Some sections are a full three feet deep. *Photo by Ray J. White*

Multiflora rose was planted along this section of the stream bank *(top photo)* in the late 1940s to protect the bank from grazing cattle. In 1973 the multiflora rose *(bottom photo)* and much additional vegetation protect the stream bank. *Top photo by Kenneth Haupt*

A long section of a trout stream bulldozed in 1972 after Hurricane Agnes. No productivity studies were made here before or after bulldozing, but most certainly there has been a significant decrease.

its success is quite limited in spite of the large amount of publicity that has been given to it.

In the above paragraph a number of environmental factors affecting the habitat were assumed to be operating as a single limiting factor, i.e., eroding banks, cows, sedimentation, acid, temperature, and so forth, whereas rarely is a single factor involved. When a correction is made often more than one condition of the environment is affected. Usually, the correction is desirable, and if the main problem is solved the secondary benefit is an unexpected bonus.

Several fundamental things must be remembered when angler conservationists set out to correct the environment. First, the time must be taken to find out what the problem is, and the ramifications of the problem must be understood insofar as that is possible. A study should be made of the stream, especially during the most adverse times of the year; the worst as well as the best should be known about the stretch of stream that is to be improved. Whatever is undertaken must be related to the biological species that live, or should live, in the water. Many stream-improvement crusades are undertaken just to do something or to get the gang together for an outing without any real knowledge of what should be done for what reason.

It should be recognized that some jobs are too big for the limited physical resources at our command. There are problems that must be attacked at the political level. Time must be spent at the capitals of the states or in Washington, D.C. We cannot solve the problems of dams, diversions, irrigation, nitrogen supersaturation, acid mine drainage in thousands of miles of water, land-use and land-management planning, the surface mining of oil-bearing shales, or the siting of nuclear power plants by building single- or double-wing walls.

We must be reasonable and realistic, well informed and knowledgeable, prepared to spend time and money on the job, and with steadfastness and determination approach the total problem and its solution as something to be done for the environment—our biosphere—not merely to catch another trout.

Epilogue

Water is never missed until the well runs dry.

Old saying.

In the preceding chapter, Dr. Alvin Grove has pointed out that a few of our waters have been rehabilitated, but the vast majority still are deteriorating. If this tendency is not reversed, soon, there will be little fishing for anybody. The older definition of a good angler was that of a man of skill who was touched by the magic of his quarry, and who achieved surcease from anxiety, and peace from being on and being surrounded by flowing water and green things. In today's world each of us must add a new dimension to our fishing. We must become dedicated guardians of those places where we fish. *All* anglers must do this now to stop and then reverse the tide of flowing water destruction.

In the first chapter, William Flick gave you the parameters of normal cold- and warmwater streams with their delicately balanced food pyramids. Ben East outlined the major forms of eco-flowing water spoilage—damming, channelization, siltation, pollution, and the criminal discarding of refuse. Don Ecker, Stan Bryer, Dave Whitlock, and Maurice Otis profiled remedial actions. Pete Van Gytenbeek suggested areas where you can go for assistance. All this gave you a good overlook, but reading this one primer is not enough. Become conscious of the conservation happenings as reported in your media, talk with informed people, read other publications. With increasing knowledge you will be able to understand what water contamination means to you, to your family, and to your community.

Conservationists can hurt their own cause by wearing blinkers and advocating too many reforms, but, on the other hand, selfish interests often present seemingly valid arguments to support their positions. Become proficient. Be able to analyze the pluses and minuses of any issue involving water.

The National Wildlife Federation, in a recent issue of their magazine, *National Wildlife*, documented a new approach to one of the basic problems concerning water—the necessary dollars to rehabilitate our brooks, streams, and rivers. In 1973, water pollution costs the head of an average American family $213 a year. If 90 percent of the present damage is corrected by 1980, the gross savings per family will be $192 per year. Deduct from these future savings the annual cost of cleanup of $105.00. Resultant savings per family by 1980 will be $87.00. This arithmetic must be meaningful to you. Money is important in this world, and polluted water is money out of your pocket.

The remedy will be expensive. But clean water is a public, social good; communities cannot survive without it. What was once free has become an essential budget cost for modern civilization.

This book has cited some major examples of desecration—instances of where you, the man who likes the out-of-doors, have been damaged. It would take ten books, no, a hundred, to detail what might be happening to your particular stream. But there is one fortunate fact; in most cases minor defilements are microcosms of bigger insults. Generally a translation is possible.

Therefore, add one more dimension to your fishing. On your home stream or anywhere else, never go on a trip without being conscious of the ecological state of the water. As you move up and down a stream, make it second nature to plot a correction survey. Is the fishing as good as it was? If not, why? Is the water getting polluted? Is the depth decreasing? Is the temperature changing? What can you do about these and other events?

If you appreciate and go angling, *get involved*. At the long-term fundamental level, suggest and support basic educational efforts. Your children must grow up imbued with a platform ethos that includes everybody's inherent right to clean water. No self-serving interests should be allowed to stand in the way.

For your own days afield, never leave a stream without planning for or physically making some improvement to the waters where you have been fishing. This can vary from water watching, calling a conservation lawyer, to planting a willow or tilting one stone to make a hide. If each fisherman cares enough, and performs one deed, picks up one can, it will help turn the tide; it will influence others, and the chain becomes endless.

If you are not yet a member, join a fishing organization; if you already belong, become more active. Many fine organizations, such as Trout Unlimited, BASS, the Federation of Fly Fishermen, the Striped Bass Fund, the several groups for saving the salmon (for addresses of these and other organizations, see the "Organizations" list) have been the instigators and have won great conservation battles.

Each of these organizations represents one fish or one type of angling. What they have accomplished is magnificent, but their effect statewide or nationally has been limited, only because individually their membership has been small. Imagine what clout an organization would have if it represented all the thirty-four million fishermen who buy freshwater licenses plus the people who fish salt water. A voice representing forty to fifty million people! An organization that speaks for one out of every six or seven persons in the United States.

Such an organization has been born. It is called the American League of Anglers.

Without detracting, but aiding what each and every local group is doing, this new group will represent in Washington, D.C., and eventually at the state capitals, all sport fishermen throughout the United States—no matter what species they fish for or how they angle. It will be supported by you, the fisherman. It will publicly lobby for your interests. It will furnish aid for your legal battles. Against defilement, it will establish uniform criteria. It will evaluate research into pollution problems and recommend the best present solutions. It will inform you, its constituency, how the battle for conservation is progressing.

Such an organization is tremendously worthwhile. All fishermen should join this and their other local groups, but this cannot be enough. Grassroots are the foundation of America,

All of man's activities must be put into the ecological perspective: that all of the natural world, including man, is interdependent, so that every act must take on a new kind of responsibility, that of safeguarding our own nature-based existence by safeguarding that of other natural forms. *Wisconsin Natural Resources Department*

and vigilant individual anglers must be the pillars for stream improvement.

Alistair Cooke recently was master of ceremonies of a TV program highlighting his version of the history of the United States. He concluded this fine series with this story:

> An Italian immigrant came to this country. For forty years he supported his family and himself by operating a shoeshine service in Grand Central, an important railroad station in New York City. He became such a fixture that a New York newspaper decided to do a feature story about him. The interview ended with the reporter asking the new citizen what he had learned about America. The man thought for a moment and then answered, "Even in America there is no free lunch."

If just one out of every ten fishermen in these United States took this story to his heart, made the following four pledges to himself and then determined to carry them out, the slide of fishing because of the deterioration of its only home—water—would stop. For you and your descendants our sport would continue. It would improve, and you, if you were one of the ten, could live with the gut feeling of having done your share for the continuance of a sport you, I, and the majority of Americans love.

Photo courtesy of Dave Whitlock

Pledges

I

Conscientiously, to make myself more knowledgeable about water—its natural state, its present contamination, and the best means of correcting the present insults.

II

To discuss with my friends, support group activities, and vote for those representatives who stand for good pollution and water controls. To do this because clean water is the inherent birthright of all human beings, and achieving it is economically sound.

III

To establish a new habit. Each and every time I go fishing, to make myself aware—not only where the biggest and most fish are, but of any ecological changes in the waters where I am fishing. To teach myself to understand those things that are harming my river, and plan what can be done to help.

IV

Never to go fishing without paying for this fun and privilege by returning some good to the waters. To do this either individually or through a group effort—because unless we all help there will be no fishing.

J. MICHAEL MIGEL

Organizations

A Selected List of Conservation Groups, Public and Private, Actively Concerned with Rivers

AMERICAN FISHERIES SOCIETY
1319 18th Street, N.W.
Washington, D.C. 20036

AMERICAN LEAGUE OF ANGLERS (ALA)
810 18th Steet, N.W.
Washington, D.C. 20006

AMERICAN RIVERS CONSERVATION COUNCIL
324 C Street, S.E.
Washington, D.C. 20003

BASS
P.O. Box 3044
Montgomery, Alabama 36109

BROTHERHOOD OF THE JUNGLE COCK
10 East Fayette Street
Baltimore, Maryland 21202

BUREAU OF LAND MANAGEMENT
United States Department of the Interior
Washington, D.C. 20240

BUREAU OF OUTDOOR RECREATION
United States Department of the Interior
Washington, D.C. 20240

BUREAU OF SPORT FISHERIES AND WILDLIFE
United States Department of the Interior
Washington, D.C. 20240

ENVIRONMENTAL DEFENSE FUND, INC.
162 Old Town Road
East Setauket, N.Y. 11733

ENVIRONMENTAL PROTECTION AGENCY
United States Waterside Mall
Washington, D.C. 20460

FEDERATION OF FLY FISHERMEN (FFF)
15513 Haas Avenue
Gardena, California 90249

HUDSON RIVER FISHERMEN'S ASSOCIATION
Lane Gate Road
Cold Spring, N.Y. 10516

IZAAK WALTON LEAGUE OF AMERICA
1800 North Kent Street
Arlington, Virginia 22209

NATIONAL MARINE FISHERIES SERVICE
3300 Whitehaven Parkway
Washington, D.C. 20240

NATIONAL PARK SERVICE
Department of the Interior
Washington, D.C. 20250

NATURAL RESOURCES DEFENSE COUNCIL
1710 N Street, N.W.
Washington, D.C. 20036

RESTORATION OF ATLANTIC SALMON IN AMERICA, INC. (RASA)
Box 164
Hancock, N.H. 03449

SCENIC HUDSON PRESERVATION CONFERENCE
545 Madison Avenue
New York, N.Y. 10022

SIERRA CLUB
1050 Mills Tower
San Francisco, California 94104

SOIL CONSERVATION SERVICE
United States Department of Agriculture
Washington, D.C. 20250

SPORT FISHING INSTITUTE
Suite 801
608 13th Street, N.W.
Washington, D.C. 20005

STRIPED BASS FUND, INC.
45–21 Glenwood Street
Little Neck, N.Y. 11362

THEODORE GORDON FLYFISHERS, INC. (TGF)
24 East 39th Street
New York, N.Y. 10016

TROUT UNLIMITED (TU)
4260 E. Evans Avenue
Denver, Colorado 80222

WATER POLLUTION CONTROL FEDERATION
3900 Wisconsin Avenue, N.W.
Washington, D.C. 20016

Note: A *Conservation Directory,* with a complete listing of prominent organizations and individuals engaged in conservation work, is available for $2.00 from:

THE NATIONAL WILDLIFE FEDERATION
1412 16th Street, N.W.
Washington, D.C. 20036

Selected Bibliography

(Books, articles, and pamphlets—collated from suggestions by the contributors)

BEHRMAN, A. S. *Water Is Everybody's Business.* Garden City, N.Y.: Doubleday & Company, Inc. 1968.

BERKMAN, RICHARD L., and VISCUSI, W. KIP. *Damning the West* (Ralph Nader's Study Group Report). New York: Bantam Books, Inc., 1972.

BOYLE, ROBERT H. *The Hudson River.* New York: W. W. Norton & Company, Inc., 1969.

CARR, DONALD E. *Death of the Sweet Waters.* New York: Berkley Publishing Corporation, 1971.

Clean Water—It's Up to You. Arlington, Virginia: The Izaak Walton League of America, n.d.

Don't Leave It All to the Experts. Washington, D.C.: United States Printing Office, n.d.

DUBOS, RENÉ. *A God Within.* New York: Charles Scribner's Sons, 1970.

EHRLICH, PAUL R. *The Population Bomb.* New York: Ballantine Books, Inc., 1972.

FLICK, ART. "Where Did the Browns Go?" *Trout* magazine Vol. 12, No. 3, Summer 1971.

GORDON, SID. *How to Fish from Top to Bottom.* Harrisburg, Pennsylvania: Telegraph Press, 1957.

HYNES, H. B. NOEL. *The Ecology of Running Water.* Toronto: The University of Toronto Press, 1972.

LANDAU, NORMAN J., and RHEINGOLD, PAUL D. *The Environmental Law Handbook.* Washington, D.C.: U.S. Government Printing Office, n.d.

MCCLANE, A. J. "Fish Culture," *McClane's Standard Fishing Encyclopaedia.* New York: Holt, Rinehart and Winston, Inc., 1965.

MACAN, T. T. *Freshwater Ecology.* New York: John Wiley & Sons, Inc., 1963.

MACKENTHUN, KENNETH M., and INGRAM, WILLIAM MARCUS. *Biological Associated Problems in Freshwater Environments, Their Identification, Investigation and Control.* Washington, D.C.: United States Printing Office, 1967.

NEEDHAM, JAMES G., and LLOYD, J. T. *The Life of Inland Waters.* Ithaca, N.Y.: Comstock Publishing Co., 1937.

NEEDHAM, PAUL R. *Trout Streams.* New York: Winchester Press, 1971.

NELSON, GAYLORD. *America's Last Chance.* Waukesha, Wisconsin: Country Beautiful Corporation, 1972.

ODUM, EUGENE P. *Ecology.* New York: Holt, Rinehart and Winston, 1970.

OTIS, MAURICE. *Guide to Stream Improvement.* New York State Conservation Department, Division of Conservation Education, n.d.

PRINGLE, LAURENCE. *Wild River.* New York: J. B. Lippincott Company, 1972.

SMITH, ROBERT L. *Ecology and Field Biology.* New York: Harper & Row, Publishers, 1966.

TALBOT, ALLAN R. *Power Along the Hudson.* New York: E. P. Dutton & Co., Inc., 1972.

UNSINGER, ROBERT L. *The Life of Rivers and Streams.* New York: McGraw-Hill Book Company, 1967.

VAN GYTENBEEK, R. P. *The Way of a Trout.* New York: J. B. Lippincott Company, 1972.

WARD, BARBARA, and DUBOS, RENÉ. *Only One Earth.* New York: Ballantine Books, Inc., 1966.

WHITE, RAY J. "Hatcheries in the Rivers," *Trout* magazine Vol. 10, No. 4, Fall 1969.

WHITLOCK, DAVE. "Trout by the Boxful," *The Flyfisher* (Parts I and II) Vol. V, Nos. 2 and 3, 1972.

ZWICK, DAVID, and BENSTOCK, MARCY. *Water Wasteland* (Ralph Nader's Study Group Report). New York: Bantam Books, Inc., 1972.

Notes on the Contributors

STANLEY BRYER, a principal in a New York City law firm, has long been active in conservation causes. He assisted in the drafting of the Wild and Scenic Rivers Bill, ultimately enacted into law by the New York State legislature, and assisted as well in obtaining important amendments to existing New York State conservation laws. He is a vice-president and the Conservation Chairman of the Theodore Gordon Flyfishers and, as general counsel for that group, has participated in administrative hearings involving the Croton River, the Beaverkill River, Schoharie Creek, and other important eastern rivers. On a national level, he has represented Trout Unlimited and the Natural Resources Defense Council in conservation actions.

BEN EAST has been a prominent and distinguished field editor for *Outdoor Life* magazine for more than forty years. Among his literally hundreds of articles have been a number of pathbreaking, dramatic pieces on pollution and desecration outrages. He is the author of *Survival* and *Danger,* coauthor of the best-selling *Silence of the North,* and a contributor to such magazines, along with *Outdoor Life,* as *National Geographic Magazine, Travel,* and *Holiday*. Ben lives in Holly, Michigan.

DON ECKER, a graduate of Harvard College, manages to spend a large part of his life involved with trout and conservation despite being a principal in an investment brokerage firm. He is outdoor editor for the *Bergen News,* the founding past president of the East Jersey Chapter of Trout Unlimited, a director of Trout Unlimited and Theodore Gordon Flyfishers, and chairman of the Conservation Commission, borough of Fort Lee.

WILLIAM FLICK's early years were spent on the streams in the Catskills where he fished and guided with his well-known father, Art. A graduate of Cornell University, Bill worked for the New York State Conservation Department for five years before joining the staff of his alma mater in 1957 as a fisheries biologist. His research on wild and hybrid strains of brook trout—carried on in the Adirondacks, where Bill lives—has been the high point of his career: his brookies live longer and grow larger than domestic strains.

ALVIN R. (BUS) GROVE is the author of *The Lure and Lore of Trout Fishing,* author or coauthor of several textbooks and many magazine and newspaper articles on angling, conservation, management of natural and renewable resources, and other topics. He is an associate dean in the College of Science at The Pennsylvania State University, editor of the magazine *Trout,* national director of Trout Unlimited, a trustee of the Museum of American Fly Fishing, and a member of the Citizens' Advisory Council of the Pennsylvania Department of Environmental Resources.

J. MICHAEL MIGEL has been for many years a successful business executive. A lifelong love of trout fishing has led him to become actively involved in important conservation, and he has published stories in *Fly Fisherman* magazine. Mike is a member of Trout Unlimited, the Theodore Gordon Flyfishers, and the Federation of Fly Fishermen, and he has been active in helping to launch the new American League of Anglers group.

MAURICE B. OTIS—whose family has served New York State conservation for more than two hundred years—has a distinguished twenty-five-year career in fish and wildlife conservation work. He has been an Engineering Technician, Assistant Supervisor of Construction, Supervisor for Lake and Stream Habitat Improvement, and Executive Assistant to the Director of Fish and Wildlife; he is currently Regional Supervisor for Fish and Wildlife in the Adirondacks. Maury's articles on stream improvement and related subjects have appeared in such magazines as the New York State *Conservationist* and the *New York State Fish and Game Journal.* He is a member of numerous conservation groups and is a past president of the Northeast Society of Conservation Engineers.

R. P. VAN GYTENBEEK, a graduate of Princeton University, is the vigorous and imaginative executive director of Trout Unlimited and author of *The Way of a Trout,* the important conservation book based on the popular TU movie of that name. Pete, who lives in Denver, is also a member of the Audubon Society and the Izaak Walton League.

DAVE WHITLOCK—an enormously gifted wildlife artist, fly-tier, and conservationist—has begun to have a major impact on the world of fly-fishing. He is responsible for introducing the Vibert Box system into the United States, he has developed a host of new fly patterns, including the Whitlock Sculpin and a series of excellent nymphs (all described in his chapter of *Art Flick's Master Fly-Tying Guide*), and his fine illustrations now grace a number of new angling books. Dave has lectured widely and given slide presentations across the country; last year he represented the United States Government in Yugoslavia, presenting demonstrations on American fly-fishing. He has published articles in *Field & Stream, Outdoor Life, Sports Afield,* and elsewhere.

Index

Pages in italics refer to illustrations.

J

K

L

M

N

O

P